British Railways Pocket Book

FORTIETH EDITION
2016

The complete guide to all Locomotive-Hauled Carriages which operate on the national railway network

Peter Hall & Robert Pritchard

PLATFORM 5

ISBN 978 1909 431 20 1

© 2015. Platform 5 Publishing Ltd, 52 Broadfield Road, Sheffield, S8 0XJ, England.

Printed in England by The Lavenham Press, Lavenham, Suffolk.

All rights reserved. No part of this publication may be reproduced in any form or transmitted in any form by any means electronic, mechanical, photocopying, recording or otherwise without the prior permission of the publisher.

CONTENTS

Provision of Information & Updates		2
Britain's Railway System		3
Introduction		10
1.	BR Number Series Stock	28
2	High Speed Train Trailer Cars	60
3.	Saloons	80
4.	Pullman Car Company Series	83
5.	Locomotive Support Coaches	86
6.	95xxx & 99xxx Range Number Conversion Table	88
7.	Set Formations	97
8.	Service Stock	100
9.	Coaching Stock Awaiting Disposal	107
10.	Codes	109

PROVISION OF INFORMATION

This book has been compiled with care to be as accurate as possible, but in some cases information is not easily available and the publisher cannot be held responsible for any errors or omissions. We would like to thank the companies and individuals who have been co-operative in supplying information to us. The authors of this series of books are always pleased to receive notification from readers of any inaccuracies readers may find, to enhance future editions. Please send comments to:

Robert Pritchard, Platform 5 Publishing Ltd, 52 Broadfield Road, Sheffield, S8 0XJ, England.　　　　　　　　**e-mail:** robert.pritchard@platform5.com
Tel: 0114 255 2625　　**Fax:** 0114 255 2471.
This book is updated to information received by 5 October 2015.

UPDATES

This book is updated to the Stock Changes given in **Today's Railways UK 167** (November 2015). Readers are therefore advised to update this book from the official Platform 5 Stock Changes published every month in **Today's Railways UK** magazine, starting with issue 168.

The Platform 5 magazine Today's Railways UK contains news and rolling stock information on the railways of Great Britain and Ireland and is published on the second Monday of every month. For further details of Today's Railways UK, please see the advertisement on the back cover of this book.

Front cover photograph: DRS Driving Open Brake Standard 9707 approaches Workington leading the 08.42 Carlisle–Barrow-in-Furness Northern service on 08/08/15.　　　　　　　　　　　　　　　　　　　　　**Robert Pritchard**

BRITAIN'S RAILWAY SYSTEM

INFRASTRUCTURE & OPERATION

Britain's national railway infrastructure is owned by a "not for dividend" company, Network Rail. In September 2014 Network Rail was classified a public sector company, being described by the Government as a "public sector arm's-length body of the Department for Transport".

Many stations and maintenance depots are leased to and operated by Train Operating Companies (TOCs), but some larger stations are under Network Rail control. The only exception is the infrastructure on the Isle of Wight: Island Line was the only franchise that included the maintenance of the infrastructure as well as the operation of passenger services. As Island Line is now part of the South West Trains franchise, both the infrastructure and trains are operated by South West Trains.

Trains are operated by TOCs over Network Rail, regulated by access agreements between the parties involved. In general, TOCs are responsible for the provision and maintenance of the locomotives, rolling stock and staff necessary for the direct operation of services, whilst Network Rail is responsible for the provision and maintenance of the infrastructure and also for staff to regulate the operation of services.

The Department for Transport is the franchising authority for the national network, with Transport Scotland overseeing the award of the ScotRail franchise. Railway Franchise holders (TOCs) can take commercial risks, although some franchises are known as "management contracts", where ticket revenues pass directly to the DfT. Concessions (such as London Overground) see the operator paid a fee to run the service, usually within tightly specified guidelines. Operators running a Concession would not normally take commercial risks, although there are usually penalties and rewards in the contract.

During 2012 the letting of new franchises was suspended pending a review of the franchise system. The process was restarted in 2013 but it is taking a number of years to catch-up and several franchises are receiving short-term extensions (or "Direct Awards") in the meantime.

DOMESTIC PASSENGER TRAIN OPERATORS

The large majority of passenger trains are operated by the TOCs on fixed-term franchises or concessions. Franchise expiry dates are shown in the list of franchisees below:

Franchise	Franchisee	Trading Name
Caledonian Sleeper	Serco (until 31 March 2030)	**Caledonian Sleeper**

This new franchise started in April 2015 when the ScotRail and ScotRail Sleeper franchises were split. Abellio won the ScotRail franchise and Serco the Caledonian Sleeper franchise. Caledonian Sleeper operates four trains nightly between London Euston and Scotland using locomotives hired from GBRf or DB Schenker. New CAF rolling stock will be introduced from 2018.

4 BRITAIN'S RAILWAY SYSTEM

Chiltern Arriva (Deutsche Bahn) **Chiltern Railways**
(until 31 December 2021)

Chiltern Railways operates a frequent service between London Marylebone, Banbury and Birmingham Snow Hill, with some peak trains extending to Kidderminster. There are also regular services from Marylebone to Stratford-upon-Avon and to Aylesbury Vale Parkway via Amersham (along the London Underground Metropolitan Line). A new route to Oxford Parkway was added to the franchise in autumn 2015, and this line will be extended to Oxford in 2016. The fleet consists of DMUs of Classes 121 (used on the Princes Risborough–Aylesbury route), 165, 168 and 172 plus a number of loco-hauled rakes used on some of the Birmingham route trains, worked by Class 68s hired from DRS.

Cross-Country Arriva (Deutsche Bahn) **CrossCountry**
(until October 2016)
Franchise extension to be negotiated to October 2019.

CrossCountry operates a network of long distance services between Scotland, North-East England and Manchester to the South-West of England, Reading, Southampton, Bournemouth and Guildford, centred on Birmingham New Street. These trains are mainly formed of diesel Class 220/221 Voyagers, supplemented by a small number of HSTs on the NE–SW route. Inter-urban services also link Nottingham, Leicester and Stansted Airport with Birmingham and Cardiff. These use Class 170 DMUs.

Crossrail MTR **Crossrail**
(until 30 May 2023)
There is an option to extend the concession by 2 years to May 2025.

This is a new concession which started in May 2015. Initially Crossrail took over the Liverpool Street–Shenfield stopping service from Abellio Greater Anglia, using a fleet of Class 315 EMUs. New Class 345 EMUs will be introduced on this route from spring 2017 and then from 2018–19 Crossrail will operate through new tunnels beneath central London, from Shenfield and Abbey Wood in the east to Reading and Heathrow in the west.

East Midlands Stagecoach Group **East Midlands Trains**
(until 4 March 2018)
There is an option to extend the franchise by 1 year to March 2019.

EMT operates a mix of long distance high speed services on the Midland Main Line (MML), from London St Pancras to Sheffield (Leeds at peak times and some extensions to York/Scarborough) and Nottingham, and local and regional services ranging from the Norwich–Liverpool route to Nottingham–Skegness, Nottingham–Mansfield–Worksop, Nottingham–Matlock and Derby–Crewe. It also operates local services in Lincolnshire. Trains on the MML are worked by a fleet of Class 222 DMUs and nine HSTs, whilst the local and regional fleet consists of DMU Classes 153, 156 and 158.

Essex Thameside National Express Group **c2c**
(until 8 November 2029)
There is an option to extend the franchise by seven reporting periods to May 2030.

c2c operates an intensive, principally commuter, service from London Fenchurch Street to Southend and Shoeburyness via both Upminster and Tilbury. The fleet consists entirely of Class 357 EMUs. In 2014 c2c won the new 15-year franchise that promised to introduce 17 new 4-car EMUs from 2019.

BRITAIN'S RAILWAY SYSTEM

Greater Western   First Group   **Great Western Railway**
(until 1 April 2019)

There is an option to extend the franchise by 1 year to April 2020.

Great Western Railway (until September 2015 branded as First Great Western) operates long distance trains from London Paddington to South Wales, the West Country and Worcester and Hereford. In addition there are frequent trains along the Thames Valley corridor to Newbury/Bedwyn and Oxford, plus local and regional trains throughout the South-West including the Cornish, Devon and Thames Valley branches, the Reading–Gatwick North Downs line and Cardiff–Portsmouth Harbour and Bristol–Weymouth regional routes. A fleet of 53 HSTs is used on the long-distance trains, with DMUs of Classes 165 and 166 used on the North Downs and Thames Valley routes and Class 180s used alongside HSTs on the Cotswold Line to Worcester and Hereford. Classes 143, 150, 153 and 158 are used on local and regional trains in the South-West. A small fleet of four Class 57s is maintained to principally work the overnight "Cornish Riviera" Sleeper service between London and Penzance.

Greater Anglia   Abellio (Netherlands Railways)   **Abellio Greater Anglia**
(until 19 October 2016)

Abellio Greater Anglia operates main line trains between London Liverpool Street, Ipswich and Norwich and local trains across Norfolk, Suffolk and parts of Cambridgeshire. It also runs local and commuter services into Liverpool Street from the Great Eastern (including Southend, Braintree and Clacton) and West Anglia (including Cambridge and Stansted Airport) routes. It operates a varied fleet of Class 90s with loco-hauled Mark 3 sets, DMUs of Classes 153, 156 and 170 and EMUs of Classes 317, 321, 360 and 379. A locomotive-hauled set, with Class 37s, is hired from DRS for use on some trains between Norwich and Great Yarmouth/Lowestoft.

Integrated Kent   Govia (Go-Ahead/Keolis)   **Southeastern**
(until 24 June 2018)

Southeastern operates all services in the South-East London suburbs, the whole of Kent and part of Sussex, which are primarily commuter services to London. It also operates domestic high speed trains on HS1 from London St Pancras to Ashford, Ramsgate, Dover and Faversham with additional peak services on other routes. EMUs of Classes 375, 376, 465 and 466 are used, along with Class 395s on the High Speed trains.

InterCity East Coast   Stagecoach/Virgin Trains   **Virgin Trains East Coast**
(until 31 March 2023)

There is an option to extend the franchise by 1 year to March 2024.

Virgin Trains East Coast operates frequent long distance trains on the East Coast Main Line between London King's Cross, Leeds, York, Newcastle and Edinburgh, with less frequent services to Bradford, Harrogate, Skipton, Hull, Lincoln, Glasgow, Aberdeen and Inverness. A mixed fleet of Class 91s and 30 Mark 4 sets, and 15 HST sets, are used on these trains.

InterCity West Coast   Virgin Rail Group (Virgin/Stagecoach Group)   **Virgin Trains**
(until September 2017)

Virgin Trains operates long distance services along the West Coast Main Line from London Euston to Birmingham/Wolverhampton, Manchester, Liverpool and Glasgow using Class 390 Pendolino EMUs. It also operates Class 221 Voyagers on the Euston–Chester–Holyhead route, whilst a mixture of 221s and 390s are used on the Euston–Birmingham–Glasgow/Edinburgh route.

BRITAIN'S RAILWAY SYSTEM

London Rail   MTR/Arriva (Deutsche Bahn)   **London Overground**
(until 12 November 2016)

This is a Concession and is different from other rail franchises, as fares and service levels are set by Transport for London instead of by the DfT.

London Overground operates services on the Richmond–Stratford North London Line and the Willesden Junction–Clapham Junction West London Line, plus the East London Line from Highbury & Islington to New Cross and New Cross Gate, with extensions to Clapham Jn (via Denmark Hill), Crystal Palace and West Croydon. It also runs services from London Euston to Watford Junction. All these use Class 378 EMUs whilst Class 172 DMUs are used on the Gospel Oak–Barking route. London Overground also took over the operation of some suburban services from London Liverpool Street in 2015 – to Chingford, Enfield Town and Cheshunt. These use Class 315 and 317 EMUs.

Merseyrail Electrics   Serco/Abellio (Netherlands Railways)   **Merseyrail**
(until 19 July 2028)

Under the control of Merseytravel PTE instead of the DfT. Franchise reviewed every five years to fit in with the Merseyside Local Transport Plan.

Merseyrail operates services between Liverpool and Southport, Ormskirk, Kirkby, Hunts Cross, New Brighton, West Kirby, Chester and Ellesmere Port, using Class 507 and 508 EMUs.

Northern Rail   Serco/Abellio (Netherlands Railways)   **Northern**
(until 1 April 2016)

Northern operates a range of inter-urban, commuter and rural services throughout the North of England, including those around the cities of Leeds, Manchester, Sheffield, Liverpool and Newcastle. The network extends from Chathill in the north to Nottingham in the south, and Cleethorpes in the east to St Bees in the west. Long distance services include Leeds–Carlisle, Middlesbrough–Carlisle and York–Blackpool North. The operator uses a large fleet of DMUs of Classes 142, 144, 150, 153, 155, 156 and 158 plus EMU Classes 319, 321, 322, 323 and 333.

ScotRail   Abellio (Netherlands Railways)   **ScotRail**
(until 31 March 2022)

There is an option to extend the franchise by 3 years to March 2025.

ScotRail provides almost all passenger services within Scotland and also trains from Glasgow to Carlisle via Dumfries, some of which extend to Newcastle (jointly operated with Northern). The company operates a large fleet of DMUs of Classes 156, 158 and 170 and EMU Classes 314, 318, 320, 334 and 380. Two loco-hauled rakes are also used on Fife Circle commuter trains, hauled by Class 68s hired from DRS.

South Western   Stagecoach Group   **South West Trains**
(until June 2017)

South West Trains operates trains from London Waterloo to destinations across the South and South-West including Woking, Basingstoke, Southampton, Portsmouth, Salisbury, Exeter, Reading and Weymouth as well as suburban services from Waterloo. SWT also runs services between Ryde and Shanklin on the Isle of Wight, using former London Underground 1938 stock (Class 483s). The rest of the fleet consists of DMU Classes 158 and 159 and EMU Classes 444, 450, 455, 456 and 458.

Thameslink & Great Northern   Govia (Go-Ahead/Keolis)   **Govia Thameslink Railway**
(until 19 September 2021)

There is an option to extend the franchise by 2 years to September 2023.

Govia operates this franchise, the largest in the UK, as a management contract. The former

BRITAIN'S RAILWAY SYSTEM

Southern franchise was combined with Thameslink/Great Northern in 2015. GTR uses four brands within the franchise: "Thameslink" operates trains between Bedford and Brighton via central London and also on the Sutton and Wimbledon loops using Class 319, 377 and 387 EMUs. Some trains continue into Southeastern territory to Sevenoaks, Orpington and Ashford. "Great Northern" comprises services from London King's Cross and Moorgate to Welwyn Garden City, Hertford North, Peterborough, Cambridge and Kings Lynn using Class 313, 317, 321 and 365 EMUs. "Southern" operates predominantly commuter services between London, Surrey and Sussex, as well as services along the South Coast between Southampton, Brighton, Hastings and Ashford, and also the cross-London service linking South Croydon and Milton Keynes. It also operates metro services in South London. Class 171 DMUs are used on Brighton–Ashford and London Bridge–Uckfield services, whilst all other services are in the hands of Class 313, 377 and 455 EMUs. Finally the premium "Gatwick Express" operates non-stop trains between London Victoria and Gatwick Airport using Class 442 EMUs (to be replaced by Class 387s in 2016). Some trains extend to/from Brighton in the peaks.

Trans-Pennine Express   First Group/Keolis   **TransPennine Express**
                                  (until 1 April 2016)

TransPennine Express operates predominantly long distance inter-urban services linking major cities across the North of England, along with Edinburgh and Glasgow in Scotland. The main services are Manchester Airport/Manchester Piccadilly–Newcastle/Middlesbrough/Hull plus Liverpool–Scarborough and Liverpool–Newcastle along the North Trans-Pennine route via Huddersfield, Leeds and York, and Manchester Airport–Cleethorpes along the South Trans-Pennine route via Sheffield. TPE also operates Manchester Airport–Blackpool/Barrow-in-Furness/Windermere/Edinburgh/Glasgow. The fleet consists of DMU Classes 170 (used on the Cleethorpes routes, but to transfer to Chiltern Railways) and 185 and also new Class 350 EMUs used on Manchester Airport–Scotland services. To cover for the loss of Class 170s to Chiltern, Class 156s are being hired from Northern for use on some Manchester Airport–Blackpool North services.

Wales & Borders   Arriva (Deutsche Bahn)   **Arriva Trains Wales**
                                  (until 14 October 2018)*

The franchise agreement includes the provision for the term to be further extended by mutual agreement by up to five years beyond October 2018. Management of the franchise is devolved to the Welsh Government, but DfT is still the procuring authority.

Arriva Trains Wales operates a mix of long distance, regional and local services throughout Wales, including the Valley Lines network of lines around Cardiff, and also through services to the English border counties and to Manchester and Birmingham. The fleet consists of DMUs of Classes 142, 143, 150, 158 and 175 and two loco-hauled rakes: one used on a premium Welsh Government sponsored service on the Cardiff–Holyhead route, and one used between Manchester/Crewe and Holyhead/Llandudno (both are hauled by a Class 67).

West Midlands   Govia (Go-Ahead/Keolis)   **London Midland**
                                  (until 31 March 2016)

Franchise extension to be negotiated to October 2017.

London Midland operates long distance and regional services from London Euston to Northampton and Birmingham/Crewe and also between Birmingham and Liverpool as well as local and regional services around Birmingham, including to Stratford-upon-Avon, Worcester, Hereford, Redditch and Shrewsbury. It also operates the Bedford–Bletchley and Watford Jn–St Albans Abbey branches. The fleet consists of DMU Classes 150, 153, 170 and 172 and EMU Classes 319, 323 and 350.

BRITAIN'S RAILWAY SYSTEM

The following operators run non-franchised services (* special summer services only):

Operator	Trading Name	Route
BAA	Heathrow Express	London Paddington–Heathrow Airport
First Hull Trains	First Hull Trains	London King's Cross–Hull
Grand Central	Grand Central	London King's Cross–Sunderland/Bradford Interchange
North Yorkshire Moors Railway Enterprises	North Yorkshire Moors Railway	Pickering–Grosmont–Whitby/Battersby
West Coast Railway Company	West Coast Railway Company	Birmingham–Stratford-upon-Avon* Fort William–Mallaig* York–Settle–Carlisle*

INTERNATIONAL PASSENGER OPERATORS

Eurostar International operates passenger services between the UK and mainland Europe. The company, established in 2010, is jointly owned by SNCF (the national operator of France): 55%, SNCB (the national operator of Belgium): 5% and Patina Rail 40%. Patina Rail is made up of Canadian-based Caisse de dépôt et placement du Québec (CDPG) and UK-based Hermes Infrastructure (each owning 30% and 10% respectively). This 40% was previously owned by the UK Government until it was sold in 2015.

In addition, a service for the conveyance of accompanied road vehicles through the Channel Tunnel is provided by the tunnel operating company, Eurotunnel.

FREIGHT TRAIN OPERATORS

The following operators operate freight services or empty passenger stock workings under "Open Access" arrangements:

Colas Rail: Colas Rail operates a number of On-Track machines and also supplies infrastructure trains for Network Rail. It also now operates a number of different freight flows, including steel, coal, oil and timber. Colas Rail has a small but varied fleet consisting of Class 37s, 47s, 56s, 60s, 66s and 70s.

DB Schenker Rail (UK): Still the biggest freight operator in the country, DBS (formerly EWS before being bought by DB) provides a large number of infrastructure trains to Network Rail and also operates coal, steel, intermodal and aggregate trains nationwide. The core fleet is Class 66s. Of the original 250 ordered 176 are still used in the UK, with the remainder having moved to DB's French and Polish operations, although some of the French locos do return to the UK when major maintenance is required. A fleet of around 20–25 Class 60s are also used on heavier trains.

DBS's six Class 59/2s are used alongside the Mendip Rail 59/0s and 59/1s on stone traffic from the Mendip quarries and around the South-East. DBS's fleet of Class 67s are used on passenger or standby duties for Arriva Trains Wales and Virgin Trains East Coast and also on excursions or special trains.

BRITAIN'S RAILWAY SYSTEM 9

Class 90s see some use on WCML freight traffic, and are also still hired to work some Caledonian Sleeper trains. The Class 92s are used on a limited number of overnight freights on High Speed 1.

DBS also operates the Class 325 EMUs for Royal Mail and a number of excursion trains.

Devon & Cornwall Railways (a subsidiary of British American Railway Services): DCRail specialises in short-term freight haulage contracts, mainly in the scrap, coal and aggregates markets, using its fleet of Class 56s. It also provides locomotives from its fleet of Class 31s or 56s for stock moves or to move On-Track Machines or other equipment.

Direct Rail Services: DRS has built on its original nuclear flask traffic to operate a number of different services. The main flows are intermodal plus the provision of crews and locos to Network Rail for autumn Railhead Treatment Trains and also operates some NR infrastructure trains. Its Class 47s, 57s and 68s are used on excursion work.

DRS has the most varied fleet of locomotives, with Class 20s, 37s, 47s, 57s and 66s working alongside a fleet of Class 68s that are currently being delivered. The company has 32 Class 68s on order as well as ten new Vossloh electric locomotives (Class 88s), that will feature a small diesel engine.

Freightliner: Freightliner has two divisions: Intermodal operates container trains from the main Ports at Southampton, Felixstowe, Tilbury and Thamesport to major cities including London, Manchester, Leeds and Birmingham. The Heavy Haul division covers the movement of coal, cement, infrastructure and aggregates nationwide. Most services are worked by Class 66s, with Class 70s mainly used on some of the heavier intermodal trains. A small fleet of Class 86 and 90 electrics is used on intermodal trains on the Great Eastern and West Coast Main Lines, the Class 86s mainly being used in pairs on the WCML between Crewe and Coatbridge. Class 90s are also hired to Caledonian Sleeper.

GB Railfreight: GBRf, owned by Eurotunnel, operates a mixture of traffic types, mainly using Class 66s together with a small fleet of Class 73s on infrastructure duties and test trains in the South-East. A growing fleet of Class 92s is also used on some intermodal flows to/from Dollands Moor or through the Channel Tunnel to Calais. Traffic includes coal, intermodal, biomass, aggregates and gypsum as well as infrastructure services for Network Rail and London Underground. GBRf also supplies locomotives to Caledonian Sleeper.

GBRf also operates some excursion trains, including those using the preserved Class 201 "Hastings" DEMU.

Rail Operations Group: This new company facilitates rolling stock movements by providing drivers or using locomotives hired from other companies.

West Coast Railway Company: WCRC has a freight licence but doesn't operate any freight as such – only empty stock movements. Its fleet of 47s, supplemented by a smaller number of 37s and 57s, is used on excursion work nationwide, including the prestigious Royal Scotsman.

In addition Amey, Balfour Beatty Rail, Harsco Rail, South West Trains, Swietelsky Babcock Rail (SB Rail) and VolkerRail operate trains formed of On-Track Machines.

ately 10

INTRODUCTION

This book contains details of all locomotive-hauled or propelled coaching stock, often referred to as carriages, which can run on Britain's national railway network.

The number of locomotive-hauled or propelled carriages in use on the national railway network is much fewer than was once the case and their number is expected to reduce further as more new multiple units are delivered. Those that remain fall into two distinct groups.

Firstly, there are those used by franchised and open access operators for regular timetabled services. Most of these are formed in fixed or semi-fixed formations with either locomotives or a locomotive and Driving Brake Carriage at either end which allows for push-pull operation. There are also a small number of mainly overnight trains with variable formations that use conventional locomotive haulage.

Secondly there are those used for what can best be described as excursion trains. These include a wide range of carriage types ranging from luxurious saloons to those more suited to the "bucket and spade" seaside type of excursion. These are formed into sets to suit the requirements of the day. From time to time some see limited use with franchised and open access operators to cover for stock shortages and times of exceptional demand such as major sporting events.

In addition there remain a small number of carriages referred to as "Service Stock" which are used internally within the railway industry and are not used to convey passengers.

FRANCHISED & OPEN ACCESS OPERATORS

For each operator regularly using locomotive-hauled carriages brief details are given here of the sphere if operation. For details of operators using HSTs see Section 2.

Abellio Greater Anglia
The Inter-City service between London Liverpool Street and Norwich is operated using 12 sets of Mark 3 carriages with Class 90 locomotives in push-pull formations. In addition one set of Mark 2 carriages hired from DRS is used between Norwich and Great Yarmouth/Lowestoft, with two Class 37s in push-pull mode.

Arriva Trains Wales
The Monday–Friday Welsh Assembly Government sponsored train between Cardiff and Holyhead uses Mark 3 carriages in push-pull mode with a Class 67. A second similarly formed set is used for weekday trains between Crewe or Manchester and Chester/North Wales. These carriages are also used for relief trains, particularly in connection with sports fixtures at Cardiff and busy ferry sailings to/from Holyhead.

Caledonian Sleeper
This franchise, operated by Serco, is a new undertaking which started in April 2015 when the Anglo-Scottish Sleeper operation was split from the

INTRODUCTION

ScotRail franchise. Caledonian Sleeper comprises seating and sleeping car services between London Euston and Scotland using sets of Mark 3 Sleeping Cars and Mark 2 seating and catering carriages. These were intended to be hauled by Class 92s between London Euston and Edinburgh/Glasgow Central but reliability problems with these locomotives has seen Class 90s substituting. Class 67s are used between Edinburgh and Aberdeen, Inverness and Fort William, but it is intended these will be replaced by Class 73/9s from late 2015.

Chiltern Railways
Chiltern operates four sets of Mark 3 carriages with Class 68 locomotives on its Mainline services between London Marylebone and Birmingham Moor Street/Kidderminster. Another set is used on a peak-hour commuter service between Marylebone and Banbury. All trains operate as push-pull sets.

Great Western Railway
The "Night Riviera" seating and sleeping car service between London Paddington and Penzance uses sets of Mark 3 carriages hauled by Class 57/6 locomotives. The seating carriages are also used in Devon and Cornwall for local services on summer Saturdays.

Northern
Northern operates two sets of Mark 2 carriages hired from DRS on the Cumbrian Coast route (between Carlisle and Barrow-in-Furness/Preston) with Class 37s in push-pull mode.

North Yorkshire Moors Railway
In addition to operating the North Yorkshire Moors Railway between Pickering and Grosmont the company operates through services to Whitby and occasionally Battersby. A fleet of Mark 1 passenger carriages and Pullman Cars are used for these services.

ScotRail
ScotRail operates two sets of Mark 2 carriages supplied by DRS on peak-hour services between Edinburgh and Fife using Class 68s.

Virgin Trains East Coast
VTEC operates 30 sets of Mark 4 carriages with Class 91 locomotives in push-pull formations on its Inter-City services between London King's Cross and Yorkshire, North-East England and Scotland.

EXCURSION TRAIN OPERATORS

Usually, three types of companies will be involved in the operation of an excursion train. There will be the promoter, the rolling stock provider and the train operator. In many cases two or more of these roles may be undertaken by the same or associated companies. Only a small number of Train Operating Companies facilitate the operation of excursion trains. This takes various forms ranging from the complete package of providing and operating the train, through offering a "hook up and haul" service, to operating the train for a third party rolling stock custodian.

DB Schenker Rail UK
DBS currently operates its own luxurious train of Mark 3 carriages, called the company train. It also offers a hook up and haul service and regularly

operates the Royal Train and the Belmond British Pullman as well as trains for Riviera Trains and its client promoters. An excursion train fleet of Mark 2 carriages is owned by DBS but these are currently not in use.

Direct Rail Services
DRS operates a small fleet of Mark 2 carriages which see occasional use on excursion trains. However, the majority are currently hired to Abellio Greater Anglia and Northern for use on regular timetabled services. Carriages are also hired to franchised operators to cover stock shortages or periods of exceptional passenger demand. The company also offers a hook up and haul service operating the Belmond Northern Belle as well as trains for Riviera Trains and their client promoters.

GB Railfreight
GBRf Initially operated excursion trains using the preserved Class 201 "Hastings" DEMU. It now also operates a small number of company excursions using hired carriages and trains for Riviera Trains and its client promoters.

West Coast Railway Company
This vertically integrated company has its own fleet of steam and diesel locomotives as well as a full range of different carriage types. It operates its own regular trains, such as the "Jacobite" steam service between Fort William and Mallaig and numerous excursion trains for itself and client promoters. In addition it offers a hook up and haul service operating the Royal Scotsman and Statesman trains as well as trains for Vintage Trains, The Princess Royal Locomotive Trust and the Scottish Railway Preservation Society.

LAYOUT OF INFORMATION

Carriages are listed in numerical order of painted number in batches according to type.

Where a carriage has been renumbered, the former number is shown in parentheses. If a carriage has been renumbered more than once, the original number is shown first in parentheses, followed by the most recent previous number.

Each carriage entry is laid out as in the following example (previous number(s) column may be omitted where not applicable):

No.	Prev. No.	Notes	Livery	Owner	Operator	Depot/Location
42346	(41053)	*h	**FD**	A	*GW*	LA

Codes: Codes are used to denote the livery, owner, operator and depot/location of each carriage. Details of codes used can be found in Section 10 of this book.

The owner is the responsible custodian of the carriage and this may not always be the legal owner.

The operator is the organisation which facilitates the use of the carriage and may not be the actual train operating company which runs the train. If no operator is shown the carriage is considered to be not in use.

The depot is the facility primarily responsible for the carriages maintenance. Light maintenance and heavy overhauls may be carried out elsewhere.

The location is where carriages not in use are currently being kept/stored.

GENERAL INFORMATION

CLASSIFICATION AND NUMBERING

Seven different numbering systems were in use on British Rail. These were the British Rail series, the four pre-nationalisation companies' series', the Pullman Car Company's series and the UIC (International Union of Railways) series. In this book BR number series carriages and former Pullman Car Company series are listed separately. There is also a separate listing of "Saloon" type carriages, that includes pre-nationalisation survivors, which are permitted to run on the national railway system, Locomotive Support Carriages and Service Stock. Please note the Mark 2 Pullman carriages were ordered after the Pullman Car Company had been nationalised and are therefore numbered in the British Rail series.

Also listed separately are the British Rail and Pullman Car Company number series carriages used on North Yorkshire Moors Railway services on the national railway system. This is due to their very restricted sphere of operation.

The BR number series grouped carriages of a particular type together in chronological order. Major modifications affecting type of accommodation resulted in renumbering into a more appropriate or new number series. Since privatisation such renumbering has not always taken place resulting in renumbering which has been more haphazard and greater variations within numbering groups.

With the introduction of the TOPS numbering system, coaching stock (including multiple unit vehicles) retained their original BR number unless this conflicted with a locomotive number. Carriages can be one–five digits, although no one or two digit examples remain in use on the national network. BR generally numbered "Service Stock" in a six-digit wagon number series.

UNITS OF MEASUREMENT

All dimensions and weights are quoted for carriages in an "as new" condition or after a major modification, such as fitting with new bogies etc. Dimensions are quoted in the order length x width. Lengths quoted are over buffers or couplers as appropriate. All widths quoted are maxima. All weights are shown as metric tonnes (t = tonnes).

DETAILED INFORMATION & CODES

Under each type heading, the following details are shown:

- "Mark" of carriage (see below).
- Descriptive text.
- Number of First Class seats, Standard Class seats, lavatory compartments and wheelchair spaces shown as F/S nT nW respectively. A number in brackets indicates tip-up seats (in addition to the regular seats).
- Bogie type (see below).
- Additional features.
- ETS Index.
- Weight: All weights are shown as metric tonnes (t = tonnes).

BOGIE TYPES

BR Mark 1 (BR1). Double bolster leaf spring bogie. Generally 90 mph, but Mark 1 bogies may be permitted to run at 100 mph with special maintenance. Weight: 6.1 t.

BR Mark 2 (BR2). Single bolster leaf-spring bogie used on certain types of non-passenger stock and suburban stock (all now withdrawn). Weight: 5.3 t.

COMMONWEALTH (C). Heavy, cast steel coil spring bogie. 100 mph. Weight: 6.75 t.

B4. Coil spring fabricated bogie. Generally 100 mph, but B4 bogies may be permitted to run at 110 mph with special maintenance. Weight: 5.2 t.

B5. Heavy duty version of B4. 100 mph. Weight: 5.3 t.

B5 (SR). A bogie originally used on Southern Region EMUs, similar in design to B5. Now also used on locomotive-hauled carriages. 100 mph.

BT10. A fabricated bogie designed for 125 mph. Air suspension.

T4. A 125 mph bogie designed by BREL (now Bombardier Transportation).

BT41. Fitted to Mark 4 carriages, designed by SIG in Switzerland. At present limited to 125 mph, but designed for 140 mph.

BRAKES

Air braking is now standard on British main line trains. Carriages with other equipment are denoted:

- b   Air braked, through vacuum pipe.
- v   Vacuum braked.
- x   Dual braked (air and vacuum).

INTRODUCTION

HEATING & VENTILATION

Electric heating and ventilation is now standard on British main-line trains. Certain carriages for use on excursion services may also have steam heating facilities, or be steam heated only. All carriages used on North Yorkshire Moors Railway trains have steam heating.

NOTES ON ELECTRIC TRAIN SUPPLY

The sum of ETS indices in a train must not be more than the ETS index of the locomotive. The normal voltage on British trains is 1000 V. Suffix "X" denotes 600 amp wiring instead of 400 amp. Trains whose ETS index is higher than 66 must be formed completely of 600 amp wired stock. Class 33 and 73/1 locomotives cannot provide a suitable electric train supply for Mark 2D, Mark 2E, Mark 2F, Mark 3, Mark 3A, Mark 3B or Mark 4 carriages. Class 55 locomotives provide an ETS directly from one of their traction generators into the train line. Consequently voltage fluctuations can result in motor-alternator flashover. Thus these locomotives are not suitable for use with Mark 2D, Mark 2E, Mark 2F, Mark 3, Mark 3A, Mark 3B or Mark 4 carriages unless modified motor-alternators are fitted. Such motor alternators were fitted to Mark 2D and 2F carriages used on the East Coast Main Line, but few remain fitted.

PUBLIC ADDRESS

It is assumed all carriages are now fitted with public address equipment, although certain stored carriages may not have this feature. In addition, it is assumed all carriages with a conductor's compartment have public address transmission facilities, as have catering carriages.

COOKING EQUIPMENT

It is assumed that Mark 1 catering carriages have gas powered cooking equipment, whilst Mark 2, 3 and 4 catering carriages have electric powered cooking equipment unless stated otherwise.

ADDITIONAL FEATURE CODES

d	Central Door Locking.
dg	Driver–Guard communication equipment.
f	Facelifted or fluorescent lighting.
h	"High density" seating
k	Composition brake blocks (instead of cast iron).
n	Day/night lighting.
pg	Public address transmission and driver-guard communication.
pt	Public address transmission facility.
q	Catering staff to shore telephone.
w	Wheelchair space.

BUILD DETAILS

Lot Numbers
Vehicles ordered under the auspices of BR were allocated a lot (batch) number when ordered and these are quoted in class headings and sub-headings.

Builders
These are shown in class headings, the following designations being used:

Ashford	BR, Ashford Works.
BRCW	Birmingham Railway Carriage & Wagon Company, Smethwick, Birmingham.
BREL Derby	BREL, Derby Carriage Works (later ABB/Adtranz Derby, now Bombardier Transportation Derby).
Charles Roberts	Charles Roberts & Company, Horbury, Wakefield (later Bombardier Transportation).
Cravens	Cravens, Sheffield.
Derby	BR, Derby Carriage Works (later BREL Derby, then ABB/Adtranz Derby, now Bombardier Transportation Derby).
Doncaster	BR, Doncaster Works (later BREL Doncaster, then BRML Doncaster, then ABB/Adtranz Doncaster, then Bombardier).
Eastleigh	BR, Eastleigh Works (later BREL Eastleigh, then Wessex Traincare, and Alstom Eastleigh, now Arlington Fleet Services).
Glasgow	BR Springburn Works, Glasgow (now Knorr-Bremse Rail Systems).
Gloucester	The Gloucester Railway Carriage & Wagon Co.
Hunslet-Barclay	Hunslet Barclay, Kilmarnock Works (now Wabtec Rail Scotland).
Metro-Cammell	Metropolitan-Cammell, Saltley, Birmingham (later GEC-Alsthom Birmingham, then Alstom Birmingham).
Pressed Steel	Pressed Steel, Linwood.
Swindon	BR Swindon Works.
Wolverton	BR Wolverton Works (later BREL Wolverton then Railcare, Wolverton, then Alstom, Wolverton now Knorr-Bremse Rail Systems).
York	BR, York Carriage Works (later BREL York, then ABB York).

Information on sub-contracting works which built parts of vehicles eg the underframes etc is not shown. In addition to the above, certain vintage Pullman cars were built or rebuilt at the following works:

Metropolitan Carriage & Wagon Company, Birmingham (later Alstom).
Midland Carriage & Wagon Company, Birmingham.
Pullman Car Company, Preston Park, Brighton.
Conversions have also been carried out at the Railway Technical Centre, Derby, LNWR, Crewe and Blakes Fabrications, Edinburgh.

▲ Pullman Car Company-liveried Kitchen Buffet Unclassified 1659 at Ravenglass on 25/07/15. **Andrew Mason**

▼ Belmond Northern Belle-liveried Mark 1 Kitchen Unclassified 1953 at Stoke Hammond (Bletchley) on 20/05/15. **Mark Beal**

▲ Royal Train Mark 2B Royal Household Couchette 2920 at Wolverton Works on 23/03/12. **Mark Beal**

▼ BR Carmine & Cream-liveried Mark 1 Open First 3123 near Fauldhouse on 13/08/15. **Robin Ralston**

▲ Riviera Trains Great Briton-liveried Mark 2F Open First 3330 at Castlethorpe on 21/06/15. **Mark Beal**

▼ West Coast Railway Company maroon-liveried Mark 1 Open Standard 4940 near Scarborough on 27/08/14. **Andrew Mason**

▲ BR Western Region Chocolate & Cream-liveried Mark 1 Open Standard 4949 at Goodrington on 02/08/15. **Robin Ralston**

▼ West Coast Railway Company Maroon-liveried Mark 2 Open Standard 5237 at Maidstone East on 29/08/15. **Robert Pritchard**

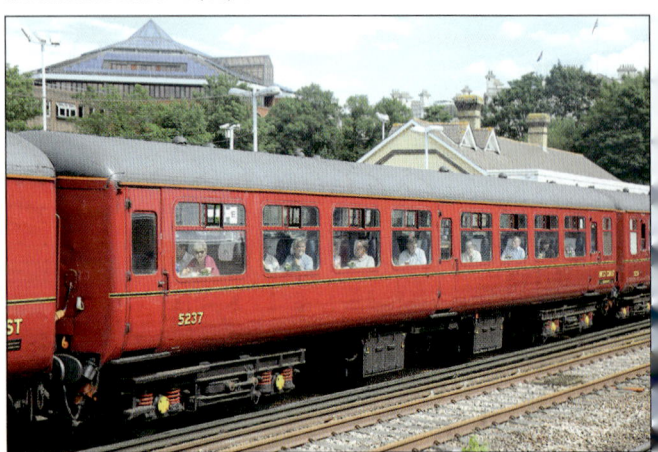

▲ Revised Direct Rail Services-liveried Mark 2F Open Standard 6117 at Norwich on 04/07/15. **Robert Pritchard**

▼ ScotRail Saltire liveried Mark 2F Open Standard 6177 at Motherwell depot on 08/04/15. **Robin Ralston**

▲ Caledonian Sleeper-liveried Mark 2E Open Brake Unclassified 9802 at Ashton on 30/05/15. **Mark Beal**

▼ Great Western Railway Green-liveried Kitchen Buffet First 10219 at Exeter St Davids on 01/08/15. **Robin Ralston**

▲ Caledonian Sleeper-liveried Mark 3A Sleeping Car with Pantry 10580 at Ashton on 30/05/15. **Mark Beal**

▼ Belmond Northern Belle-liveried Mark 3A Sleeping Car 10734 near Carstairs on 15/08/15. **Robin Ralston**

▲ Abellio Greater Anglia-liveried Mark 3A Open Standard 12019 at Diss on 05/07/15. **Robert Pritchard**

▼ Virgin Trains East Coast-liveried Mark 4 Open Standard 12478 at Doncaster on 28/05/15. **Robert Pritchard**

ABBREVIATIONS

The following abbreviations are used in class headings and also throughout this publication:

BR	British Railways.
DB	Deutsche Bahn
DEMU	Diesel Electric Multiple Unit.
DMU	Diesel Multiple Unit (general term).
EMU	Electric Multiple Unit.
ETH	Electric Train Heating
ft	feet
GWR	Great Western Railway
kN	kilonewtons.
km/h	kilometres per hour.
kW	kilowatts.
LT	London Transport.
LUL	London Underground Limited.
m	metres.
mph	miles per hour.
SR	BR Southern Region
t	tonnes.

26 THE DEVELOPMENT OF BR STANDARD COACHES

THE DEVELOPMENT OF BR STANDARD COACHES

Mark 1

The standard BR coach built from 1951 to 1963 was the Mark 1. This type features a separate underframe and body. The underframe is normally 64 ft 6 in long, but certain vehicles were built on shorter (57 ft) frames. Tungsten lighting was standard and until 1961, BR Mark 1 bogies were generally provided. In 1959 Lot No. 30525 (Open Standard) appeared with fluorescent lighting and melamine interior panels, and from 1961 onwards Commonwealth bogies were fitted in an attempt to improve the quality of ride which became very poor when the tyre profiles on the wheels of the BR1 bogies became worn. Later batches of Open Standard and Open Brake Standard retained the features of Lot No. 30525, but compartment vehicles – whilst utilising melamine panelling in Standard Class – still retained tungsten lighting. Wooden interior finish was retained in First Class vehicles where the only change was to fluorescent lighting in open vehicles (except Lot No. 30648, which had tungsten lighting). In later years many Mark 1 coaches had BR 1 bogies replaced by B4. More recently a small number of carriages have had BR1 bogies replaced with Commonwealth bogies.

XP64

In 1964, a new prototype train was introduced. Known as "XP64", it featured new seat designs, pressure heating & ventilation, aluminium compartment doors and corridor partitions, foot pedal operated toilets and B4 bogies. The vehicles were built on standard Mark 1 underframes. Folding exterior doors were fitted, but these proved troublesome and were later replaced with hinged doors. All XP64 coaches have been withdrawn, but some have been preserved.

Mark 2

The prototype Mark 2 vehicle (W13252) was produced in 1963. This was a Corridor First of semi-integral construction and had pressure heating & ventilation, tungsten lighting, and was mounted on B4 bogies. This vehicle has now been preserved at the Mid Norfolk Railway. The production build was similar, but wider windows were used. The Open Standard vehicles used a new seat design similar to that in the XP64 and fluorescent lighting was provided. Interior finish reverted to wood. Mark 2 vehicles were built from 1964–66.

Mark 2A–2C

The Mark 2A design, built 1967–68, incorporated the remainder of the features first used in the XP64 coaches, ie foot pedal operated toilets (except Open Brake Standard), new First Class seat design, aluminium compartment doors and partitions together with fluorescent lighting in first class compartments. Folding gangway doors (lime green coloured) were used instead of the traditional one-piece variety.

THE DEVELOPMENT OF BR STANDARD COACHES 27

Mark 2B coaches had wide wrap around doors at vehicle ends, no centre doors and a slightly longer body. In Standard Class there was one toilet at each end instead of two at one end as previously. The folding gangway doors were red.

Mark 2C coaches had a lowered ceiling with twin strips of fluorescent lighting and ducting for air conditioning, but air conditioning was never fitted.

Mark 2D–2F

These vehicles were fitted with air conditioning. They had no opening toplights in saloon windows, which were shallower than previous ones.

Mark 2E vehicles had smaller toilets with luggage racks opposite. The folding gangway doors were fawn coloured.

Mark 2F vehicles had a modified air conditioning system, plastic interior panels and InterCity 70 type seats.

Mark 3

The Mark 3 design has BT10 bogies, is 75 ft (23 m) long and is of fully integral construction with InterCity 70 type seats. Gangway doors were yellow (red in Kitchen Buffet First) when new, although these were changed on refurbishment. Locomotive-hauled coaches are classified Mark 3A, Mark 3 being reserved for HST trailers. A new batch of Open First and Open Brake First, classified Mark 3B, was built in 1985 with Advanced Passenger Train-style seating and revised lighting. The last vehicles in the Mark 3 series were the driving brake vans ("Driving Van Trailers") built for West Coast Main Line services but now used elsewhere.

A number of Mark 3 vehicles have recently been converted for use as HST trailers with CrossCountry, Grand Central and Great Western Railway.

Mark 4

The Mark 4 design was built by Metro-Cammell for use on the East Coast Main Line after electrification and featured a body profile suitable for tilting trains, although tilt is not fitted, and is not intended to be. This design is suitable for 140 mph running, although is restricted to 125 mph because the signalling system on the route is not suitable for the higher speed. The bogies for these coaches were built by SIG in Switzerland and are designated BT41. Power operated sliding plug exterior doors are standard. All Mark 4s were rebuilt with completely new interiors in 2003–05 for GNER and referred to as "Mallard" stock. These rakes generally run in fixed formations and are now operated by Virgin Trains East Coast.

1. BR NUMBER SERIES COACHING STOCK

KITCHEN FIRST

Mark 1. Spent most of its life as a Royal Train vehicle and was numbered 2907 for a time. 24/-. B5 bogies. ETS 2.

Lot No. 30633 Swindon 1961. 41 t.

325 **VN** BE *NB* CP DUART

PULLMAN KITCHEN

Mark 2. Pressure Ventilated. Built with First Class seating but this has been replaced with a servery area. Gas cooking. 2T. B5 bogies. ETS 6.

Lot No. 30755 Derby 1966. 40 t.

504	**PC**	WC	*WC*	CS	ULLSWATER
506	**PC**	WC	*WC*	CS	WINDERMERE

PULLMAN OPEN FIRST

Mark 2. Pressure Ventilated. 36/- 2T. B4 bogies. ETS 5.

Lot No. 30754 Derby 1966. 35 t.

Non-standard livery: 546 Maroon & beige.

546	**O**	WC		CS	CITY OF MANCHESTER
548	**PC**	WC	*WC*	CS	GRASMERE
549	**PC**	WC	*WC*	CS	BASSENTHWAITE
550	**PC**	WC	*WC*	CS	RYDAL WATER
551	**PC**	WC	*WC*	CS	BUTTERMERE
552	**PC**	WC	*WC*	CS	ENNERDALE WATER
553	**PC**	WC	*WC*	CS	CRUMMOCK WATER

PULLMAN OPEN BRAKE FIRST

Mark 2. Pressure Ventilated. 30/- 2T. B4 bogies. ETS 4.

Lot No. 30753 Derby 1966. 35 t.

586 **PC** WC *WC* CS DERWENTWATER

1200–1730

BUFFET FIRST

Mark 2F. Air conditioned. Converted 1988–89/91 at BREL, Derby from Mark 2F Open Firsts. 1200/01/03/11/20/21 have Stones equipment, others have Temperature Ltd. 25/– 1T 1W. B4 bogies. d. ETS 6X.

1200/03/11/20. Lot No. 30845 Derby 1973. 33 t.
1201/07/10/12/21/54. Lot No. 30859 Derby 1973–74. 33 t.

1200	(3287, 6459)	**RV**	RV	*RV*	EH	AMBER
1201	(3361, 6445)	**CH**	VT	*WC*	CS	
1203	(3291)	**IC**	RV	*RV*	EH	
1207	(3328, 6422)	**V**	BE		ZG	
1210	(3405, 6462)	**FS**	E	*CA*	PO	
1211	(3305)	**PC**	RF	*ST*	CS	
1212	(3427, 6453)	**V**	RV	*RV*	EH	
1220	(3315, 6432)	**FS**	E	*CA*	PO	
1221	(3371)	**IC**	BE		ZG	
1254	(3391)	**BG**	DR		BH	

KITCHEN WITH BAR

Mark 1. Built with no seats but three Pullman-style seats now fitted in bar area. B5 bogies. ETS 1.

Lot No. 30624 Cravens 1960–61. 41 t.

1566	**VN**	BE	*NB*	CP

KITCHEN BUFFET UNCLASSIFIED

Mark 1. Built with 23 loose chairs. All remaining vehicles refurbished with 23 fixed polypropylene chairs and fluorescent lighting. 1683/91/92 were further refurbished with 21 chairs, wheelchair space and carpets. ETS 2 (* 2X).

Now used on excursion trains with the seating area adapted to various uses including servery and food preparation areas, with some or all seating removed.

1651–92. Lot No. 30628 Pressed Steel 1960–61. Commonwealth bogies. 39 t.
1730. Lot No. 30512 BRCW 1960–61. B5 bogies. 37 t.

1651		**CC**	RV	*RV*	EH		1683	**RB**	RV		BU
1657		**CH**	RV		ZG		1691	**CC**	RV	*RV*	EH
1659		**PC**	RF	*ST*	CS		1692	**CH**	RV		EH
1666	x	**M**	RP	*WC*	CS		1730	x **M**	SP	*SP*	BO
1671	x*	**CH**	RV	*RV*	EH						

BUFFET STANDARD

Mark 1. These carriages are basically an open standard with two full window spaces removed to accommodate a buffet counter, and four seats removed to allow for a stock cupboard. All remaining vehicles now have fluorescent lighting. –/44 2T. Commonwealth bogies. ETS 3.

1861 has had its toilets replaced with store cupboards.

1813–32. Lot No. 30520 Wolverton 1960. 38 t.
1840. Lot No. 30507 Wolverton 1960. 37 t.
1859–63. Lot No. 30670 Wolverton 1961–62. 38 t.
1882. Lot No. 30702 Wolverton 1962. 38 t.

1813	x	**CH**	RV	*RV*	EH		1860	x	**M**	WC	*WC*	CS
1832	x	**CC**	RV	*RV*	EH		1861	x	**M**	WC	*WC*	CS
1840	v	**M**	WC	*WC*	CS		1863	x	**CH**	LS		CM
1859	x	**M**	SP	*SP*	BO		1882	x	**M**	WC	*WC*	CS

KITCHEN UNCLASSIFIED

Mark 1. These carriages were built as Unclassified Restaurants. They were rebuilt with buffet counters and 23 fixed polypropylene chairs, then further refurbished by fitting fluorescent lighting. Further modified for use as servery vehicle with seating removed and kitchen extended. ETS 2X.

1953. Lot No. 30575 Swindon 1960. B4/B5 bogies. 36.5 t.
1961. Lot No. 30632 Swindon 1961. Commonwealth bogies. 39 t.

1953		**VN**	BE	*NB*	CP		1961	x	**M**	WC	*WC*	CS

HM THE QUEEN'S SALOON

Mark 3. Converted from an Open First built 1972. Consists of a lounge, bedroom and bathroom for HM The Queen, and a combined bedroom and bathroom for the Queen's dresser. One entrance vestibule has double doors. Air conditioned. BT10 bogies. ETS 9X.

Lot No. 30886 Wolverton 1977. 36 t.

2903	(11001)	**RP**	NR	*RT*	ZN

HRH THE DUKE OF EDINBURGH'S SALOON

Mark 3. Converted from an Open Standard built 1972. Consists of a combined lounge/dining room, a bedroom and a shower room for the Duke, a kitchen and a valet's bedroom and bathroom. Air conditioned. BT10 bogies. ETS 15X.

Lot No. 30887 Wolverton 1977. 36 t.

2904	(12001)	**RP**	NR	*RT*	ZN

ROYAL HOUSEHOLD SLEEPING CAR

Mark 3A. Built to similar specification as Sleeping Cars 10647–729. 12 sleeping compartments for use of Royal Household with a fixed lower berth and a hinged upper berth. 2T plus shower room. Air conditioned. BT10 bogies. ETS 11X.

Lot No. 31002 Derby/Wolverton 1985. 44 t.

2915   **RP** NR *RT*   ZN

HRH THE PRINCE OF WALES'S DINING CAR

Mark 3. Converted from HST TRUK built 1976. Large kitchen retained, but dining area modified for Royal use seating up to 14 at central table(s). Air conditioned. BT10 bogies. ETS 13X.

Lot No. 31059 Wolverton 1988. 43 t.

2916 (40512)   **RP** NR *RT*   ZN

ROYAL KITCHEN/HOUSEHOLD DINING CAR

Mark 3. Converted from HST TRUK built 1977. Large kitchen retained and dining area slightly modified with seating for 22 Royal Household members. Air conditioned. BT10 bogies. ETS 13X.

Lot No. 31084 Wolverton 1990. 43 t.

2917 (40514)   **RP** NR *RT*   ZN

ROYAL HOUSEHOLD CARS

Mark 3. Converted from HST TRUKs built 1976/77. Air conditioned. BT10 bogies. ETS 10X.

Lot Nos. 31083 (* 31085) Wolverton 1989. 41.05 t.

2918 (40515)   **RP** NR   ZN
2919 (40518) *   **RP** NR   ZN

ROYAL HOUSEHOLD COUCHETTES

Mark 2B. Converted from Corridor Brake First built 1969. Consists of luggage accommodation, guard's compartment, workshop area, 350 kW diesel generator and staff sleeping accommodation. B5 bogies. ETS 2X.

Lot No. 31044 Wolverton 1986. 48 t.

2920 (14109, 17109)   **RP** NR *RT*   ZN

Mark 2B. Converted from Corridor Brake First built 1969. Consists of luggage accommodation, kitchen, brake control equipment and staff accommodation. B5 bogies. ETS 7X.

Lot No. 31086 Wolverton 1990. 41.5 t.

2921 (14107, 17107) **RP** NR *RT* ZN

HRH THE PRINCE OF WALES'S SLEEPING CAR

Mark 3B. Air conditioned. BT10 bogies. ETS 7X.

Lot No. 31035 Derby/Wolverton 1987.

2922 **RP** NR *RT* ZN

ROYAL SALOON

Mark 3B. Air conditioned. BT10 bogies. ETS 6X.

Lot No. 31036 Derby/Wolverton 1987.

2923 **RP** NR *RT* ZN

OPEN FIRST

Mark 1. 42/– 2T. ETS 3. Many now fitted with table lamps.

3058 was numbered DB 975313, 3068 was numbered DB 975606 and 3093 was numbered DB 977594 for a time when in departmental service for BR.

3058–69. Lot No. 30169 Doncaster 1955. B4 bogies. 33 t (* Commonwealth bogies 35 t).
3093. Lot No. 30472 BRCW 1959. B4 bogies. 33 t.
3096–3100. Lot No. 30576 BRCW 1959. B4 bogies. 33 t.

3058 *x **M**	WC *WC*	CS	FLORENCE	3096 x **M**	SP *SP*	BT
3066	**CC**	RV *RV*	EH	3097	**CC**	RV *RV*	EH
3068	**CC**	RV *RV*	EH	3098 x **CH**	RV *RV*	EH
3069	**CC**	RV *RV*	EH	3100 x **CH**	RV *RV*	EH
3093 x **M**	WC *WC*	CS	FLORENCE			

Later design with fluorescent lighting, aluminium window frames and Commonwealth bogies.

3128/36/41/43/44/46/47/48 were renumbered 1058/60/63/65/66/68/69/70 when reclassified Restaurant Open First, then 3600/05/08/09/02/06/04/10 when declassified to Open Standard, but have since regained their original numbers. 3136 was numbered DB 977970 for a time when in use with Serco Railtest as a Brake Force Runner.

3105 has had its luggage racks removed and has tungsten lighting.

3105–28. Lot No. 30697 Swindon 1962–63. 36 t.
3130–50. Lot No. 30717 Swindon 1963. 36 t.

3105–3275　　　　　　　　　　　　　　　　　　　　　　　　　　　　33

3105	x	**M**	WC	*WC*	CS	3128	x	**M**	WC	*WC*	CS
3106	x	**M**	WC	*WC*	CS	3130	x	**M**	WC	*WC*	CS
3107	x	**CH**	LS		CL	3133	x	**M**	RV		EH
3110	x	**CH**	RV	*RV*	EH	3136	x	**M**	WC	*WC*	CS
3112	x	**CH**	RV	*RV*	EH	3140	x	**CH**	LS		CL
3113	x	**M**	WC	*WC*	CS	3141		**M**	RV		BU
3115	x	**M**	SP	*SP*	BT	3143	x	**M**	WC	*WC*	CS
3117	x	**M**	WC	*WC*	CS	3144	x	**M**	RV		BQ
3119		**CC**	RV	*RV*	EH	3146		**M**	RV		EH
3120		**CC**	RV	*RV*	EH	3147		**CH**	RV	*RV*	EH
3121		**CH**	RV	*RV*	EH	3148		**CC**	LS		CL
3122	x	**CH**	RV	*RV*	EH	3149		**CH**	RV	*RV*	EH
3123		**CC**	RV	*RV*	EH	3150		**M**	SP	*SP*	BO
3125	x	**CH**	LS		CL						

Names:

3105	JULIA		3128	VICTORIA
3106	ALEXANDRA		3130	PAMELA
3113	JESSICA		3136	DIANA
3117	CHRISTINA		3143	PATRICIA

OPEN FIRST

Mark 2D. Air conditioned. Stones equipment. 42/– 2T. B4 bogies. ETS 5.

† Interior modified to Pullman Car standards with new seating, new panelling, tungsten lighting and table lights for the Belmond Northern Belle.

Lot No. 30821 Derby 1971–72. 34 t.

3174	†	**VN**	BE	*NB*	CP	GLAMIS
3182	†	**VN**	BE	*NB*	CP	WARWICK
3188		**PC**	RF	*ST*	CS	CADAIR IDRIS

OPEN FIRST

Mark 2E. Air conditioned. Stones equipment. 42/– 2T (p 36/– 2T). B4 bogies. ETS 5.

r Refurbished with new seats.
† Interior modified to Pullman Car standards with new seating, new panelling, tungsten lighting and table lights for the Belmond Northern Belle.

Lot No. 30843 Derby 1972–73. 32.5 t. († 35.8 t).

3223		**RV**	RF		BU	DIAMOND
3231	p	**PC**	RF	*ST*	CS	BEN CRUACHAN
3232	dr	**BG**	BE		CL	
3240		**RV**	RF		BU	SAPPHIRE
3247	†	**VN**	BE	*NB*	CP	CHATSWORTH
3267	†	**VN**	BE	*NB*	CP	BELVOIR
3273	†	**VN**	BE	*NB*	CP	ALNWICK
3275	†	**VN**	BE	*NB*	CP	HARLECH

OPEN FIRST

Mark 2F. Air conditioned. 3277–3318/3358–79 have Stones equipment, others have Temperature Ltd. All refurbished in the 1980s with power-operated vestibule doors, new panels and new seat trim. 42/– 2T. B4 bogies. d. ETS 5X.

r Further refurbished with table lamps and modified seats with burgundy seat trim.
u Fitted with power supply for Mark 1 Kitchen Buffet Unclassified.

3277–3318. Lot No. 30845 Derby 1973. 33.5 t.
3325–3426. Lot No. 30859 Derby 1973–74. 33.5 t.
3431–3438. Lot No. 30873 Derby 1974–75. 33.5 t.

3277		**AR**	RV		CY	3352	r	**M**	WC	*WC*	CS
3278	r	**BP**	RV	*RV*	EH	3356	r	**RV**	RV	*RV*	EH
3279	u	**M**	DB		MH	3358		**M**	DB		MH
3292		**M**	DB		MH	3359	r	**M**	WC	*WC*	CS
3295		**AR**	RV		CY	3360	r	**PC**	WC	*WC*	CS
3304	r	**V**	RV	*RV*	EH	3362	r	**PC**	WC	*WC*	CS
3312		**PC**	RF	*ST*	CS	3364	r	**RV**	RV	*RV*	EH
3313	r	**M**	WC	*WC*	CS	3366	r	**BG**	DR		ZG
3314	r	**V**	RV		BU	3374		**BG**	DR		BH
3318		**M**	DB		MH	3379	u	**AR**	RV		CY
3325	r	**V**	RV	*RV*	EH	3384	r	**RV**	RV	*RV*	EH
3326	r	**M**	WC	*WC*	CS	3386	r	**V**	RV	*RV*	EH
3330	r	**RV**	RV	*RV*	EH	3390	r	**RV**	RV	*RV*	EH
3331		**M**	DB		MH	3392	r	**M**	WC	*WC*	CS
3333	r	**V**	RV	*RV*	EH	3395	r	**M**	WC	*WC*	CS
3334		**AR**	RV		CY	3397	r	**RV**	RV	*RV*	EH
3336	u	**AR**	RV		EH	3400		**M**	DB		MH
3340	r	**V**	RV	*RV*	EH	3417		**AR**	RV		BU
3344	r	**V**	RV	*RV*	EH	3424		**M**	DB		MH
3345	r	**V**	RV	*RV*	EH	3426	r	**RV**	RV	*RV*	EH
3348	r	**RV**	RV	*RV*	EH	3431	r	**M**	WC	*WC*	CS
3350	r	**M**	WC	*WC*	CS	3438	r	**PC**	RF	*ST*	CS
3351		**CH**	VT	*WC*	CS						

Names:

3312	HELVELLYN		3384	DICKENS
3330	BRUNEL		3390	CONSTABLE
3348	GAINSBOROUGH		3397	WORDSWORTH
3356	TENNYSON		3426	ELGAR
3364	SHAKESPEARE		3438	BEN LOMOND

OPEN STANDARD

Mark 1. –/64 2T. ETS 4.

4831–36. Lot No. 30506 Wolverton 1959. Commonwealth bogies. 33 t.
4856. Lot No. 30525 Wolverton 1959–60. B4 bogies. 33 t.

4831	x	**M**	SP	*SP*	BO	4836	x	**M**	SP	*SP*	BO
4832	x	**M**	SP	*SP*	BO	4856	x	**M**	SP	*SP*	BO

OPEN STANDARD

Mark 1. Commonwealth bogies. –/64 2T. ETS 4.

4905. Lot No. 30646 Wolverton 1961. 36 t.
4927–5044. Lot No. 30690 Wolverton 1961–62. 37 t.

4905	x	**M**	WC	*WC*	CS	4991		**CH**	RV	*RV*	EH
4927	x	**CC**	RV	*RV*	EH	4994	x	**M**	WC	*WC*	CS
4931	v	**M**	WC	*WC*	CS	4998		**CH**	RV	*RV*	EH
4940	x	**M**	WC	*WC*	CS	5007		**G**	RV		EH
4946	x	**CH**	RV	*RV*	EH	5008	x	**M**	RV		EH
4949	x	**CH**	RV	*RV*	EH	5009	x	**CH**	RV		EH
4951	x	**M**	WC	*WC*	CS	5027		**G**	RV		EH
4954	v	**M**	WC	*WC*	CS	5028	x	**M**	SP	*SP*	BO
4959		**CH**	RV	*RV*	EH	5032	x	**M**	WC	*WC*	CS
4960	x	**M**	WC	*WC*	CS	5033	x	**M**	WC	*WC*	CS
4973	x	**M**	WC	*WC*	CS	5035	x	**M**	WC	*WC*	CS
4984	x	**M**	WC	*WC*	CS	5044	x	**M**	WC	*WC*	CS

OPEN STANDARD

Mark 2. Pressure ventilated. –/64 2T. B4 bogies. ETS 4.

Lot No. 30751 Derby 1965–67. 32 t.

5157	v	**CH**	VT	*VT*	TM	5200	v	**M**	WC	*WC*	CS
5171	v	**M**	WC	*WC*	CS	5212	v	**CH**	VT	*VT*	TM
5177	v	**CH**	VT	*VT*	TM	5216	v	**M**	WC	*WC*	CS
5191	v	**CH**	VT	*VT*	TM	5222	v	**M**	WC	*WC*	CS
5198	v	**CH**	VT	*VT*	TM						

OPEN STANDARD

Mark 2. Pressure ventilated. –/48 2T. B4 bogies. ETS 4.

Lot No. 30752 Derby 1966. 32 t.

5229		**M**	WC	*WC*	CS	5239		**M**	WC	*WC*	CS
5236	v	**M**	WC	*WC*	CS	5249	v	**M**	WC	*WC*	CS
5237	v	**M**	WC	*WC*	CS						

OPEN STANDARD

Mark 2A. Pressure ventilated. –/64 2T (w –/62 2T). B4 bogies. ETS 4.

f Facelifted vehicles.

5276–5341. Lot No. 30776 Derby 1967–68. 32 t.
5366–5419. Lot No. 30787 Derby 1968. 32 t.

5276	f	**RV**	RV		BQ	5341	f	**CC**	RV	*RV*	EH
5278		**M**	WC	*WC*	CS	5366	f	**CH**	RV	*RV*	EH
5292	f	**CC**	RV	*RV*	EH	5419	w	**M**	WC	*WC*	CS

OPEN STANDARD

Mark 2B. Pressure ventilated. –/62. B4 bogies. ETS 4.

Lot No. 30791 Derby 1969. 32 t.

5487   **M** WC *WC* CS

OPEN STANDARD

Mark 2D. Air conditioned. Stones equipment. Refurbished with new seats and end luggage stacks. –/58 2T. B4 bogies. d. ETS 5.

Lot No. 30822 Derby 1971. 33 t.

5631	**M**	DB	MH	5700	**FP**	DR	ZG
5632	**M**	DB	MH	5710	**FP**	DR	BH
5657	**M**	DB	MH				

OPEN STANDARD

Mark 2E. Air conditioned. Stones equipment. –/64 2T. B4 bogies. d. ETS 5.

r Refurbished with new interior panelling.
s Refurbished with new interior panelling, modified design of seat headrest and centre luggage stack. –/60 2T.
w Fitted with one wheelchair space. –/62 2T 1W.

5787–97. Lot No. 30837 Derby 1972. 33.5 t.
5810. Lot No. 30844 Derby 1972–73. 33.5 t.

| 5787 | s | **V** | DR | | BH | 5810 | w | **DS** | DR | *NO* | KM |
| 5797 | r | **IC** | RF | | BU | | | | | | |

OPEN STANDARD

Mark 2F. Air conditioned. Temperature Ltd equipment. InterCity 70 seats. All were refurbished in the 1980s with power-operated vestibule doors, new panels and seat trim. –/64 2T. B4 bogies. d. ETS 5X.

* Early Mark 2 style seats. These vehicles have undergone a second refurbishment with carpets and new seat trim.

5910–6310 37

q Fitted with two wheelchair spaces. –/60 2T 2W.
s Fitted with centre luggage stack. –/60 2T.
t Fitted with centre luggage stack and wheelchair space. –/58 2T 1W.
w Fitted with one wheelchair space. –/62 2T 1W.

5910–55. Lot No. 30846 Derby 1973. 33 t.
5959–6158. Lot No. 30860 Derby 1973–74. 33 t.
6173–83. Lot No. 30874 Derby 1974–75. 33 t.

5910	q	**V**	RV		CY	6008		**DS**	DR	*NO*	KM
5912		**PC**	RF	*ST*	CS	6012		**M**	WC	*WC*	CS
5919	s pt	**DS**	DR	*GA*	KM	6021		**M**	WC	*WC*	CS
5921		**AR**	RV	*RV*	EH	6022	s	**M**	WC	*WC*	CS
5922		**M**	DB		BS	6024	s	**V**	RV	*RV*	EH
5924		**M**	DB		BS	6027	q	**SR**	RV	*SR*	ML
5928		**CH**	VT	*WC*	CS	6036	*	**M**	DB		BS
5929		**AR**	RV	*RV*	EH	6042		**AR**	RV	*RV*	EH
5937		**V**	RV		ZG	6046		**DS**	DR	*GA*	KM
5945		**SR**	RV	*SR*	ML	6051		**V**	RV	*RV*	EH
5950		**AR**	RV	*RV*	EH	6054		**V**	RV	*RV*	EH
5952		**SR**	RV	*SR*	ML	6064	s	**DS**	DR	*NO*	KM
5954		**M**	DB		MH	6067	s pt	**V**	RV	*RV*	EH
5955		**SR**	RV	*SR*	ML	6103		**M**	WC	*WC*	CS
5959	n	**M**	DB		BS	6110		**M**	DB		MH
5961	s pt	**V**	RV	*RV*	EH	6115	s	**M**	WC	*WC*	CS
5964		**AR**	RV	*RV*	EH	6117		**DS**	DR	*GA*	KM
5965	t	**SR**	RV	*SR*	ML	6122		**DS**	DR	*NO*	KM
5971		**DS**	DR	*NO*	KM	6137	s pt	**SR**	RV	*SR*	ML
5976	t	**SR**	RV	*SR*	ML	6139	n*	**M**	DB		MH
5985		**AR**	RV	*RV*	EH	6141	q	**V**	RV		CY
5987		**SR**	RV	*SR*	ML	6152	*	**M**	DB		BS
5991		**PC**	RF	*ST*	CS	6158		**V**	RV	*RV*	EH
5995		**DS**	DR	*NO*	KM	6173	w	**DS**	DR	*NO*	KM
5998		**AR**	RV	*RV*	EH	6176	t	**SR**	RV	*SR*	ML
6000	t	**M**	WC	*WC*	CS	6177	s	**SR**	RV	*SR*	ML
6001		**DS**	DR	*NO*	KM	6183	s	**SR**	RV	*SR*	ML
6006		**AR**	RV		CY						

BRAKE GENERATOR VAN

Mark 1. Renumbered 1989 from BR departmental series. Converted from Gangwayed Brake Van in 1973 to three-phase supply brake generator van for use with HST trailers. Modified 1999 for use with loco-hauled stock. B5 bogies.

Lot No. 30400 Pressed Steel 1958.

6310 (81448, 975325) **CH** RV *RV* EH

GENERATOR VAN

Mark 1. Converted from Gangwayed Brake Vans in 1992. B4 bogies. ETS 75.

6311. Lot No. 30162 Pressed Steel 1958. 37.25 t.
6312. Lot No. 30224 Cravens 1956. 37.25 t.
6313. Lot No. 30484 Pressed Steel 1958. 37.25 t.

6311	(80903, 92911)	**B**	DB		TO
6312	(81023, 92925)	**M**	WC	*WC*	CS
6313	(81553, 92167)	**PC**	BE	*BP*	SL

BUFFET STANDARD

Mark 2C. Converted from Open Standard by removal of one seating bay and replacing this by a counter with a space for a trolley, now replaced with a more substantial buffet. Adjacent toilet removed and converted to steward's washing area/store. Pressure ventilated. –/55 1T. B4 bogies. ETS 4.

Lot No. 30795 Derby 1969–70. 32.5 t.

6528	(5592)	**M**	WC	*WC*	CS

SLEEPER RECEPTION CAR

Mark 2F. Converted from Open First. These vehicles consist of pantry, microwave cooking facilities, seating area for passengers (with loose chairs, staff toilet plus two bars). Now refurbished again with new "sofa" seating as well as the loose chairs. Converted at RTC, Derby (6700), Ilford (6701–05) and Derby (6706–08). Air conditioned. 6700/01/03/05–08 have Stones equipment and 6702/04 have Temperature Ltd equipment. The number of seats per coach can vary but typically is 25/– 1T (12 seats as "sofa" seating and 13 loose chairs). B4 bogies. d. ETS 5X.

6700–02/04/08. Lot No. 30859 Derby 1973–74. 33.5 t.
6703/05–07. Lot No. 30845 Derby 1973. 33.5 t.

6700	(3347)	**FS**	E	*CA*	PO
6701	(3346)	**CA**	E	*CA*	PO
6702	(3421)	**FS**	E	*CA*	PO
6703	(3308)	**FS**	E	*CA*	PO
6704	(3341)	**FS**	E	*CA*	PO
6705	(3310, 6430)	**FS**	E	*CA*	PO
6706	(3283, 6421)	**FS**	E	*CA*	PO
6707	(3276, 6418)	**FS**	E	*CA*	PO
6708	(3370)	**FS**	E	*CA*	PO

6722–9494   39

BUFFET FIRST

Mark 2D. Converted from Buffet Standard by the removal of another seating bay and fitting a more substantial buffet counter with boiler and microwave oven. Now converted to First Class with new seating and end luggage stacks. Air conditioned. Stones equipment. 30/– 1T. B4 bogies. d. ETS 5.
Lot No. 30822 Derby 1971. 33 t.

6722	(5736, 6661)	**FP**	E		LM
6723	(5641, 6662)	**M**	WC		CS
6724	(5721, 6665)	**M**	WC	*WC*	CS

OPEN BRAKE STANDARD WITH TROLLEY SPACE

Mark 2. This vehicle uses the same bodyshell as the Mark 2 Corridor Brake First. Converted from Open Brake Standard by removal of one seating bay and replacing this with a counter with a space for a trolley. Adjacent toilet removed and converted to a steward's washing area/store. –/23. B4 bogies. ETS 4.

Lot No. 30757 Derby 1966. 31 t.

| 9101 | (9398) | v | **CH** | VT | *VT* | | TM |

OPEN BRAKE STANDARD

Mark 2. These vehicles use the same body shell as the Mark 2 Corridor Brake First and have First Class seat spacing and wider tables. Pressure ventilated. –/31 1T. B4 bogies. ETS 4.

9104 was originally numbered 9401. It was renumbered when converted to Open Brake Standard with trolley space. Now returned to original layout.

Lot No. 30757 Derby 1966. 31.5 t.

| 9104 | v | **M** | WC | *WC* | CS | | 9392 | v | **M** | WC | *WC* | CS |
| 9391 | | **M** | WC | *WC* | CS | | | | | | | |

OPEN BRAKE STANDARD

Mark 2D. Air conditioned. Stones Equipment. All now refurbished with new seating –/22 1TD. B4 bogies. d. pg. ETS 5.

Lot No. 30824 Derby 1971. 33 t.

| 9488 | **SR** | DR | | ZG | | 9494 | **M** | DB | MH |
| 9493 | **M** | WC | *WC* | CS | | | | | |

OPEN BRAKE STANDARD

Mark 2E. Air conditioned. Stones Equipment. –/32 1T. B4 bogies. d. pg. ETS 5.
Lot No. 30838 Derby 1972. 33 t.

Non-standard livery: 9502 Pullman umber & cream.

r Refurbished with new interior panelling.
s Refurbished with modified design of seat headrest and new interior panelling.

9496	r	**M**	VT	*WC*	CS	9507	s	**V**	RV	*RV*	EH
9502	s	**O**	BE	*BP*	SL	9508	s	**BG**	DR		BH
9504	s	**V**	RV		CY	9509	s	**AV**	RV		BU
9506	s	**BG**	DR		MH						

OPEN BRAKE STANDARD

Mark 2F. Air conditioned. Temperature Ltd equipment. All were refurbished in the 1980s with power-operated vestibule doors, new panels and seat trim. All now further refurbished with carpets. –/32 1T. B4 bogies. d. pg. ETS 5X.

9537 has had all its seats removed for the purpose of carrying luggage.

Advertising livery: 9537 CruiseSaver Express (dark blue).

Lot No. 30861 Derby 1974. 34 t.

9520	n	**AR**	RV	*RV*	EH	9527	n	**AR**	RV	*NO*	KM
9521		**DS**	RV	*SR*	ML	9529	n	**M**	DB		BS
9522		**M**	DB		MH	9531		**M**	DB		BS
9525		**DS**	DR	*GA*	KM	9537	n	**AL**	RV		EH
9526	n	**IC**	RV	*RV*	EH	9539		**SR**	RV	*SR*	ML

DRIVING OPEN BRAKE STANDARD

Mark 2F. Air conditioned. Temperature Ltd equipment. Push & pull (tdm system). Converted from Open Brake Standard, these vehicles originally had half cabs at the brake end. They have since been refurbished and have had their cabs widened and the cab-end gangways removed. Five vehicles (9701–03/08/14) have been converted for use in Network Rail test trains and can be found in the Service Stock section of this book. –/30 1W 1T. B4 bogies. d. pg. Cowcatchers. ETS 5X.

9704–10. Lot No. 30861 Derby 1974. Converted Glasgow 1979. Disc brakes. 34 t.
9711/13. Lot No. 30861 Derby 1974. Converted Glasgow 1985. 34 t.

9704	(9512)	**AR**	DR		ZA	9710	(9518)	**1**	DR		ZA
9705	(9519)	**DS**	DR	*NO*	KM	9711	(9532)	**AR**	VT		TM
9707	(9511)	**DS**	DR	*NO*	KM	9713	(9535)	**AR**	DR		ZA
9709	(9515)	**AR**	DR		ZA						

9800–10259

OPEN BRAKE UNCLASSIFIED

Mark 2E. Converted from Open Standard with new seating for use on Anglo-Scottish overnight services by Railcare, Wolverton. Air conditioned. Stones equipment. –/31 2T. B4 bogies. d. ETS 4X.

9801–03. Lot No. 30837 Derby 1972. 33.5 t.
9804–10. Lot No. 30844 Derby 1972–73. 33.5 t.

9800	(5751)	**FS**	E	*CA*	PO	9806	(5840)	**FS**	E	*CA*	PO
9801	(5760)	**FS**	E	*CA*	PO	9807	(5851)	**FS**	E	*CA*	PO
9802	(5772)	**CA**	E	*CA*	PO	9808	(5871)	**FS**	E	*CA*	PO
9803	(5799)	**FS**	E	*CA*	PO	9809	(5890)	**FS**	E	*CA*	PO
9804	(5826)	**FS**	E	*CA*	PO	9810	(5892)	**FS**	E	*CA*	PO
9805	(5833)	**FS**	E	*CA*	PO						

KITCHEN BUFFET FIRST

Mark 3A. Air conditioned. Converted from HST catering vehicles and Mark 3 Open Firsts. Refurbished with table lamps and burgundy seat trim (except *). 18/– plus two seats for staff use (* 24/–, † 24/–, § 24/–, t 23/– 1W). BT10 bogies. d. ETS 14X.

§ First Great Western Sleeper "day coaches" that have been fitted with former HST First Class seats to a 2+1 layout.

Non-standard livery: 10211 EWS dark maroon.

10200–211. Lot No. 30884 Derby 1977. 39.8 t.
10212–229. Lot No. 30878 Derby 1975–76. 39.8 t.
10232–259. Lot No. 30890 Derby 1979. 39.8 t.

10200	(40519)	*	**1**	P		*GA*	NC	10229	(11059)	*	**GA**	P	*GA*	NC
10202	(40504)	†	**BG**	AV			LM	10232	(10027)	§	**FD**	P	*GW*	PZ
10203	(40506)	*	**GA**	P	*GA*	NC	10233	(10013)		**V**	AV		LM
10211	(40510)		**O**	DB	*DB*	TO	10235	(10015)	†	**BG**	AV		LM
10212	(11049)		**VT**	P	*GA*	NC	10237	(10022)		**DR**	AV		LM
10214	(11034)	*	**GA**	P	*GA*	NC	10241	(10009)	*	**1**	P		IL
10215	(11032)		**BG**	AV		LM	10242	(10002)		**BG**	AV		LM
10216	(11041)	*	**GA**	P	*GA*	NC	10246	(10014)	†	**BG**	AV		LM
10217	(11051)		**VT**	P		ZN	10247	(10011)	*	**GA**	P	*GA*	NC
10219	(11047)	§	**GW**	P	*GW*	PZ	10249	(10012)		**AW**	AV	*AW*	CF
10222	(11063)		**BG**	AV		LM	10250	(10020)		**V**	AV		LM
10225	(11014)	§	**GW**	P	*GW*	PZ	10257	(10007)	†	**BG**	AV		LM
10226	(11015)		**V**	AV		LM	10259	(10025)	t	**AW**	AV	*AW*	CF
10228	(11035)	*	**GA**	P	*GA*	NC							

KITCHEN BUFFET FIRST

Mark 3A. Air conditioned. Rebuilt for Chiltern Railways 2011–12 and fitted with sliding plug doors. Interiors originally refurbished for Wrexham & Shropshire with Primarius seating, a new kitchen area and universal-access toilet. 30/– 1TD 1W. BT10 bogies. ETS 14X.

10271/273/274. Lot No. 30890 Derby 1979. 41.3 t.
10272. Lot No. 30884 Derby 1977. 41.3 t.

10271	(10018, 10236)	**CM** AV	*CR*	AL
10272	(40517, 10208)	**CM** AV	*CR*	AL
10273	(10021, 10230)	**CM** AV	*CR*	AL
10274	(10010, 10255)	**CM** AV	*CR*	AL

KITCHEN BUFFET STANDARD

Mark 4. Air conditioned. Rebuilt from First to Standard Class with bar adjacent to seating area instead of adjacent to end of coach. –/30 1T. BT41 bogies. ETS 6X.

Lot No. 31045 Metro-Cammell 1989–92. 43.2 t.

10300	**VE**	E	*VE*	BN	10317	**VE**	E	*VE*	BN
10301	**VE**	E	*VE*	BN	10318	**VE**	E	*VE*	BN
10302	**EC**	E	*VE*	BN	10319	**VE**	E	*VE*	BN
10303	**VE**	E	*VE*	BN	10320	**VE**	E	*VE*	BN
10304	**VE**	E	*VE*	BN	10321	**VE**	E	*VE*	BN
10305	**VE**	E	*VE*	BN	10323	**EC**	E	*VE*	BN
10306	**VE**	E	*VE*	BN	10324	**VE**	E	*VE*	BN
10307	**VE**	E	*VE*	BN	10325	**VE**	E	*VE*	BN
10308	**VE**	E	*VE*	BN	10326	**VE**	E	*VE*	BN
10309	**VE**	E	*VE*	BN	10328	**VE**	E	*VE*	BN
10310	**VE**	E	*VE*	BN	10329	**VE**	E	*VE*	BN
10311	**VE**	E	*VE*	BN	10330	**VE**	E	*VE*	BN
10312	**VE**	E	*VE*	BN	10331	**EC**	E	*VE*	BN
10313	**VE**	E	*VE*	BN	10332	**EC**	E	*VE*	BN
10315	**VE**	E	*VE*	BN	10333	**VE**	E	*VE*	BN

BUFFET STANDARD

Mark 3A. Air conditioned. Converted from Mark 3 Open Standards at Derby 2006. –/52 1T (including 6 Compin Pegasus seats for "priority" use). s (refurbished): –/54. BT10 bogies. d. ETS 13X.

Lot No. 30877 Derby 1975–77. 37.8 t.

10401	(12168)	s	**GA**	P	*GA*	NC	10404	(12068)		**GA** P	*GA* NC
10402	(12010)	1	P	*GA*	NC		10405	(12157)	s	**GA** P	*GA* NC
10403	(12135)	1	P	*GA*	NC		10406	(12020)		**GA** P	*GA* NC

BUFFET STANDARD

Mark 3A. Air conditioned. Converted from Mark 3 Kitchen Buffet First 2015–16 for Abellio Greater Anglia. Full details awaited.

10411	(, 102)		P		
10412	(, 102)		P		
10413	(, 102)		P		
10414	(, 102)		P		
10415	(11043, 10223)	**GA**	P	*GA*	NC

10416–10683 43

10416	(, 102)		P
10417	(, 102)		P

SLEEPING CAR WITH PANTRY

Mark 3A. Air conditioned. Retention toilets. 12 compartments with a fixed lower berth and a hinged upper berth, plus an attendant's compartment. 2T. BT10 bogies. d. ETS 7X.

Non-standard livery: 10546 EWS dark maroon.

Lot No. 30960 Derby 1981–83. 41 t.

10501	**FS**	P	*CA*	PO		10553	**FS**	P	*CA*	PO
10502	**FS**	P	*CA*	PO		10561	**FS**	P	*CA*	PO
10504	**FS**	P	*CA*	PO		10562	**FS**	P	*CA*	PO
10506	**FS**	P	*CA*	PO		10563	**FD**	P	*GW*	PZ
10507	**FS**	P	*CA*	PO		10565	**FS**	P	*CA*	PO
10508	**FS**	P	*CA*	PO		10580	**CA**	P	*CA*	PO
10513	**FS**	P	*CA*	PO		10584	**GW**	P	*GW*	PZ
10516	**FS**	P	*CA*	PO		10589	**FD**	P	*GW*	PZ
10519	**FS**	P	*CA*	PO		10590	**FD**	P	*GW*	PZ
10520	**FS**	P	*CA*	PO		10594	**GW**	P	*GW*	PZ
10522	**FS**	P	*CA*	PO		10596	**GW**	P		ZN
10523	**FS**	P	*CA*	PO		10597	**FS**	P	*CA*	PO
10526	**FS**	P	*CA*	PO		10598	**FS**	P	*CA*	PO
10527	**FS**	P	*CA*	PO		10600	**FS**	P	*CA*	PO
10529	**FS**	P	*CA*	PO		10601	**GW**	P	*GW*	PZ
10531	**FS**	P	*CA*	PO		10605	**FS**	P	*CA*	PO
10532	**GW**	P	*GW*	PZ		10607	**FS**	P	*CA*	PO
10534	**FD**	P	*GW*	PZ		10610	**FS**	P	*CA*	PO
10542	**FS**	P	*CA*	PO		10612	**FD**	P	*GW*	PZ
10543	**FS**	P	*CA*	PO		10613	**FS**	P	*CA*	PO
10544	**FS**	P	*CA*	PO		10614	**FS**	P	*CA*	PO
10546	**0**	DB	*DB*	TO		10616	**FD**	P	*GW*	PZ
10548	**FS**	P	*CA*	PO		10617	**FS**	P	*CA*	PO
10551	**FS**	P	*CA*	PO						

SLEEPING CAR

Mark 3A. Air conditioned. Retention toilets. 13 compartments with a fixed lower berth and a hinged upper berth (* 11 compartments with a fixed lower berth and a hinged upper berth + one compartment for a disabled person. 1TD). 2T. BT10 bogies. ETS 6X.

10734 was originally 2914 and used as a Royal Train staff sleeping car. It has 12 berths and a shower room and is ETS 11X.

10648–729. Lot No. 30961 Derby 1980–84. 43.5 t.
10734. Lot No. 31002 Derby/Wolverton 1985. 42.5 t.

10648	d*	**FS**	P	*CA*	PO		10675 d	**FS**	P	*CA*	PO
10650	d*	**FS**	P	*CA*	PO		10680 d*	**FS**	P	*CA*	PO
10666	d*	**FS**	P	*CA*	PO		10683 d	**FS**	P	*CA*	PO

10688 d	**FS**	P	*CA*	PO	10714 d*	**FS**	P	*CA*	PO
10689 d*	**FS**	P	*CA*	PO	10718 d*	**FS**	P	*CA*	PO
10690	**FS**	P	*CA*	PO	10719 d*	**FS**	P	*CA*	PO
10693 d	**CA**	P	*CA*	PO	10722 d*	**FS**	P	*CA*	PO
10699 d*	**FS**	P	*CA*	PO	10723 d*	**FS**	P	*CA*	PO
10703 d	**FS**	P	*CA*	PO	10729	**VN**	BE	*NB*	CP
10706 d*	**FS**	P	*CA*	PO	10734	**VN**	BE	*NB*	CP

Names:

10729 CREWE | 10734 BALMORAL

OPEN FIRST

Mark 3A. Air conditioned. All refurbished with table lamps and new seat cushions and trim. 48/– 2T (* 48/– 1T 1TD). BT10 bogies. d. ETS 6X.

† Reseated with Standard Class seats: –/68 2T 2W.

11006/007 were open composites 11906/907 for a time.

Non-standard livery: 11039 EWS dark maroon.

Lot No. 30878 Derby 1975–76. 34.3 t.

11006	**V**	DR		BH	11029 †	**BG**	AV	*CR*	AL
11007	**VT**	P	*GA*	NC	11031 †	**BG**	AV	*CR*	AL
11011 *	**V**	DR		BH	11033	**DR**	AV		LM
11018	**VT**	P	*GA*	NC	11039	**0**	DB	*DB*	TO
11028	**V**	AV		ZB	11048	**VT**	P	*GA*	NC

OPEN FIRST

Mark 3B. Air conditioned. InterCity 80 seats. All refurbished with table lamps and new seat cushions and trim. 48/– 2T. BT10 bogies. d. ETS 6X.

† Abellio Greater Anglia vehicles fitted with disabled toilet and reduced seating including three Compin "Pegasus" seats of the same type as used in Standard Class (but class as First Class). 37/– 1T 1TD 2W.

s Refurbished Abellio Greater Anglia vehicles (toilets reduced to one per coach). 48/– 1T or † 37/– 1TD 2W.

Lot No. 30982 Derby 1985. 36.5 t.

11066	**GA**	P	*GA*	NC	11079	**V**	AV		LM
11067 s	**GA**	P	*GA*	NC	11080 s	**GA**	P	*GA*	NC
11068	**1**	P	*GA*	NC	11081	**1**	P	*GA*	NC
11069	**GA**	P	*GA*	NC	11082	**GA**	P	*GA*	NC
11070	**GA**	P	*GA*	NC	11085 †	**1**	P	*GA*	NC
11072	**GA**	P	*GA*	NC	11087 †	**GA**	P	*GA*	NC
11073	**1**	P	*GA*	NC	11088	**GA**	P	*GA*	NC
11075	**GA**	P	*GA*	NC	11090 †	**GA**	P	*GA*	NC
11076	**GA**	P	*GA*	NC	11091	**GA**	P	*GA*	NC
11077	**GA**	P	*GA*	NC	11092 †	**GA**	P	*GA*	NC
11078 †	**GA**	P	*GA*	NC	11093 †s	**GA**	P	*GA*	NC

11094	†	**1**	P	*GA*	NC	11098	†	**1**	P	*GA*	NC
11095	†	**GA**	P	*GA*	NC	11099	†s	**GA**	P	*GA*	NC
11096	†s	**GA**	P	*GA*	NC	11100	†	**1**	P	*GA*	NC
11097		**V**	AV		LM	11101	†	**1**	P	*GA*	NC

OPEN FIRST

Mark 4. Air conditioned. Rebuilt with new interior by Bombardier Wakefield 2003–05 (some converted from Standard Class vehicles) 41/– 1T (plus 2 seats for staff use). BT41 bogies. ETS 6X.

11201–11273. Lot No. 31046 Metro-Cammell 1989–92. 41.3 t.
11277–11299. Lot No. 31049 Metro-Cammell 1989–92. 41.3 t.

11201		**EC**	E	*VE*	BN	11284	(12487)	**VE**	E	*VE*	BN
11219		**VE**	E	*VE*	BN	11285	(12537)	**VE**	E	*VE*	BN
11229		**VE**	E	*VE*	BN	11286	(12482)	**VE**	E	*VE*	BN
11237		**VE**	E	*VE*	BN	11287	(12527)	**VE**	E	*VE*	BN
11241		**VE**	E	*VE*	BN	11288	(12517)	**VE**	E	*VE*	BN
11244		**VE**	E	*VE*	BN	11289	(12528)	**VE**	E	*VE*	BN
11273		**VE**	E	*VE*	BN	11290	(12530)	**VE**	E	*VE*	BN
11277	(12408)	**VE**	E	*VE*	BN	11291	(12535)	**VE**	E	*VE*	BN
11278	(12479)	**VE**	E	*VE*	BN	11292	(12451)	**VE**	E	*VE*	BN
11279	(12521)	**VE**	E	*VE*	BN	11293	(12536)	**VE**	E	*VE*	BN
11280	(12523)	**EC**	E	*VE*	BN	11294	(12529)	**VE**	E	*VE*	BN
11281	(12418)	**VE**	E	*VE*	BN	11295	(12475)	**VE**	E	*VE*	BN
11282	(12524)	**EC**	E	*VE*	BN	11298	(12416)	**VE**	E	*VE*	BN
11283	(12435)	**VE**	E	*VE*	BN	11299	(12532)	**EC**	E	*VE*	BN

OPEN FIRST (DISABLED)

Mark 4. Air conditioned. Rebuilt from Open First by Bombardier Wakefield 2003–05. 42/– 1W 1TD. BT41 bogies. ETS 6X.

Lot No. 31046 Metro-Cammell 1989–92. 40.7 t.

11301	(11215)	**VE**	E	*VE*	BN	11316	(11227)	**VE**	E	*VE*	BN
11302	(11203)	**EC**	E	*VE*	BN	11317	(11223)	**VE**	E	*VE*	BN
11303	(11211)	**VE**	E	*VE*	BN	11318	(11251)	**VE**	E	*VE*	BN
11304	(11257)	**VE**	E	*VE*	BN	11319	(11247)	**VE**	E	*VE*	BN
11305	(11261)	**VE**	E	*VE*	BN	11320	(11255)	**VE**	E	*VE*	BN
11306	(11276)	**VE**	E	*VE*	BN	11321	(11245)	**VE**	E	*VE*	BN
11307	(11217)	**EC**	E	*VE*	BN	11322	(11228)	**VE**	E	*VE*	BN
11308	(11263)	**VE**	E	*VE*	BN	11323	(11235)	**VE**	E	*VE*	BN
11309	(11259)	**VE**	E	*VE*	BN	11324	(11253)	**VE**	E	*VE*	BN
11310	(11272)	**EC**	E	*VE*	BN	11325	(11231)	**VE**	E	*VE*	BN
11311	(11221)	**VE**	E	*VE*	BN	11326	(11206)	**VE**	E	*VE*	BN
11312	(11225)	**VE**	E	*VE*	BN	11327	(11236)	**VE**	E	*VE*	BN
11313	(11210)	**VE**	E	*VE*	BN	11328	(11274)	**VE**	E	*VE*	BN
11314	(11207)	**EC**	E	*VE*	BN	11329	(11243)	**VE**	E	*VE*	BN
11315	(11238)	**VE**	E	*VE*	BN	11330	(11249)	**VE**	E	*VE*	BN

OPEN FIRST

Mark 4. Air conditioned. Rebuilt from Open First by Bombardier Wakefield 2003–05. Separate area for 7 smokers, although smoking is no longer allowed. 46/– 1W 1TD. BT41 bogies. ETS 6X.

Lot No. 31046 Metro-Cammell 1989–92. 42.1 t.

11401	(11214)	**VE**	E	*VE*	BN	11416	(11254) **VE** E *VE* BN
11402	(11216)	**EC**	E	*VE*	BN	11417	(11226) **VE** E *VE* BN
11403	(11258)	**VE**	E	*VE*	BN	11418	(11222) **VE** E *VE* BN
11404	(11202)	**VE**	E	*VE*	BN	11419	(11250) **VE** E *VE* BN
11405	(11204)	**EC**	E	*VE*	BN	11420	(11242) **VE** E *VE* BN
11406	(11205)	**VE**	E	*VE*	BN	11421	(11220) **VE** E *VE* BN
11407	(11256)	**EC**	E	*VE*	BN	11422	(11232) **VE** E *VE* BN
11408	(11218)	**VE**	E	*VE*	BN	11423	(11230) **VE** E *VE* BN
11409	(11262)	**VE**	E	*VE*	BN	11424	(11239) **VE** E *VE* BN
11410	(11260)	**EC**	E	*VE*	BN	11425	(11234) **VE** E *VE* BN
11411	(11240)	**VE**	E	*VE*	BN	11426	(11252) **VE** E *VE* BN
11412	(11209)	**VE**	E	*VE*	BN	11427	(11200) **VE** E *VE* BN
11413	(11212)	**VE**	E	*VE*	BN	11428	(11233) **VE** E *VE* BN
11414	(11246)	**EC**	E	*VE*	BN	11429	(11275) **VE** E *VE* BN
11415	(11208)	**VE**	E	*VE*	BN	11430	(11248) **VE** E *VE* BN

OPEN FIRST

Mark 4. Air conditioned. Converted from Kitchen Buffet Standard with new interior by Bombardier Wakefield 2005. 46/– 1T. BT41 bogies. ETS 6X.

Lot No. 31046 Metro-Cammell 1989–92. 41.3 t.

11998 (10314) **VE** E *VE* BN | 11999 (10316) **VE** E *VE* BN

OPEN STANDARD

Mark 3A. Air conditioned. All refurbished with modified seat backs and new layout and further refurbished with new seat trim. –/76 2T († –/70 2T 1W, t –/72 2T, z –/70 1TD 1T 2W. BT10 bogies. d. ETS 6X.

h Abellio Greater Anglia modified coaches with 8 Compin Pegasus seats at saloon ends for "priority" use and more unidirectional seating. –/80 2T.

s Refurbished Abellio Greater Anglia coaches with more unidirectional seating and one toilet removed. –/80 1T.

§ First Great Western Sleeper "day coaches" that have been fitted with former HST First Class seats to a 2+1 layout and are effectively unclassified. –/45(2) 2T 1W.

12170/171 were converted from Open Composites 11909/910, formerly Open Firsts 11009/010.

Non-standard livery: 12142 GWR brown (trial livery).

12005–167. Lot No. 30877 Derby 1975–77. 34.3 t.
12170/171. Lot No. 30878 Derby 1975–76. 34.3 t.

12005–12171

12005 h	**1**	P	*GA*	NC	12093 h	**1**	P	*GA*	NC
12009 s	**GA**	P	*GA*	NC	12094	**V**	AV	*CR*	AL
12011	**VT**	P	*GA*	NC	12097 h	**1**	P	*GA*	NC
12012 h	**GA**	P	*GA*	NC	12098	**1**	P	*GA*	NC
12013	**GA**	P	*GA*	NC	12099 h	**GA**	P	*GA*	NC
12015 s	**GA**	P	*GA*	NC	12100 §	**FD**	P	*GW*	PZ
12016	**1**	P	*GA*	NC	12103	**1**	P	*GA*	NC
12017 t	**BG**	AV	*CR*	AL	12104	**V**	AV		LM
12019 h	**GA**	P	*GA*	NC	12105 h	**GA**	P	*GA*	NC
12021	**GA**	P	*GA*	NC	12107 h	**1**	P	*GA*	NC
12024 h	**1**	P	*GA*	NC	12108	**GA**	P	*GA*	NC
12026 h	**GA**	P	*GA*	NC	12109 h	**GA**	P	*GA*	NC
12027 s	**GA**	P	*GA*	NC	12110 h	**GA**	P	*GA*	NC
12030	**GA**	P	*GA*	NC	12111	**1**	P	*GA*	NC
12031	**GA**	P	*GA*	NC	12114 h	**GA**	P	*GA*	NC
12032 h	**1**	P	*GA*	NC	12115 h	**1**	P	*GA*	NC
12034	**1**	P	*GA*	NC	12116 h	**GA**	P	*GA*	NC
12035 h	**GA**	P	*GA*	NC	12118	**GA**	P	*GA*	NC
12037 h	**GA**	P	*GA*	NC	12119 t	**BG**	AV	*CR*	AL
12040 h	**GA**	P	*GA*	NC	12120 s	**GA**	P	*GA*	NC
12041 h	**GA**	P	*GA*	NC	12122 z	**VT**	P	*GA*	NC
12042 h	**1**	P	*GA*	NC	12125 h	**GA**	P	*GA*	NC
12043 †	**BG**	AV	*CR*	AL	12126 s	**GA**	P	*GA*	NC
12046 s	**GA**	P	*GA*	NC	12129 h	**GA**	P	*GA*	NC
12047 z	**V**	DR		BH	12130	**1**	P	*GA*	NC
12049	**1**	P	*GA*	NC	12132	**GA**	P	*GA*	NC
12051 h	**GA**	P	*GA*	NC	12133	**VT**	P	*GA*	NC
12054 †	**BG**	AV	*CR*	AL	12134	**V**	DR		BH
12056 h	**GA**	P	*GA*	NC	12137	**1**	P	*GA*	NC
12057 h	**GA**	P	*GA*	NC	12138	**VT**	P	*GA*	NC
12058	**V**	AV		LM	12139	**GA**	P	*GA*	NC
12060	**1**	P	*GA*	NC	12141	**GA**	P	*GA*	NC
12061 h	**GA**	P	*GA*	NC	12142 z	**0**	P		ZN
12062 h	**GA**	P	*GA*	NC	12143	**1**	P	*GA*	NC
12063	**1**	DR		BH	12146	**GA**	P	*GA*	NC
12064	**GA**	P	*GA*	NC	12147 s	**GA**	P	*GA*	NC
12065	**1**	DR		BH	12148 s	**GA**	P	*GA*	NC
12066 h	**1**	P	*GA*	NC	12150 h	**GA**	P	*GA*	NC
12067	**GA**	P	*GA*	NC	12151	**1**	P	*GA*	NC
12073 h	**GA**	P	*GA*	NC	12153 s	**GA**	P	*GA*	NC
12078	**VT**	P	*GA*	NC	12154 h	**GA**	P	*GA*	NC
12079	**1**	P	*GA*	NC	12159 s	**GA**	P	*GA*	NC
12081	**1**	P	*GA*	NC	12161 §	**FD**	P	*GW*	PZ
12082 s	**GA**	P	*GA*	NC	12164	**1**	P	*GA*	NC
12084 h	**1**	P	*GA*	NC	12165	**V**	AV		LM
12087 †	**V**	DR		BH	12166	**1**	P	*GA*	NC
12089	**GA**	P	*GA*	NC	12167 h	**GA**	P	*GA*	NC
12090 s	**GA**	P	*GA*	NC	12170 s	**GA**	P	*GA*	NC
12091 s	**GA**	P	*GA*	NC	12171	**GA**	P	*GA*	NC

OPEN STANDARD

Mark 3A (†) or Mark 3B. Air conditioned. BT10 bogies. d. ETS 6X. Converted from Mark 3A or 3B Open Firsts for Arriva Trains Wales. Fitted with new Grammer seating. –/70 2T 1W.

12176–181/185. Mark 3B. Lot No. 30982 Derby 1985. 38.5 t.
12182–184. Mark 3A. Lot No. 30878 Derby 1975–76. 38.5 t.

12176	(11064)	**AW** AV	*AW*	CF
12177	(11065)	**AW** AV	*AW*	CF
12178	(11071)	**AW** AV	*AW*	CF
12179	(11083)	**AW** AV	*AW*	CF
12180	(11084)	**AW** AV	*AW*	CF
12181	(11086)	**AW** AV	*AW*	CF
12182	(11013) †	**AW** AV	*AW*	CF
12183	(11027) †	**AW** AV	*AW*	CF
12184	(11044) †	**AW** AV	*AW*	CF
12185	(11089)	**AW** AV	*AW*	CF

OPEN STANDARD (END)

Mark 4. Air conditioned. Rebuilt with new interior by Bombardier Wakefield 2003–05. Separate area for 26 smokers, although smoking is no longer allowed. –/76 1T. BT41 bogies. ETS 6X.

12232 was converted from the original 12405.

12200–231. Lot No. 31047 Metro-Cammell 1989–91. 39.5 t.
12232. Lot No. 31049 Metro-Cammell 1989–92. 39.5 t.

12200	**VE**	E	*VE*	BN	12217	**VE**	E	*VE*	BN
12201	**VE**	E	*VE*	BN	12218	**VE**	E	*VE*	BN
12202	**VE**	E	*VE*	BN	12219	**VE**	E	*VE*	BN
12203	**VE**	E	*VE*	BN	12220	**VE**	E	*VE*	BN
12204	**VE**	E	*VE*	BN	12222	**VE**	E	*VE*	BN
12205	**VE**	E	*VE*	BN	12223	**VE**	E	*VE*	BN
12207	**VE**	E	*VE*	BN	12224	**VE**	E	*VE*	BN
12208	**VE**	E	*VE*	BN	12225	**VE**	E	*VE*	BN
12209	**VE**	E	*VE*	BN	12226	**VE**	E	*VE*	BN
12210	**VE**	E	*VE*	BN	12227	**VE**	E	*VE*	BN
12211	**VE**	E	*VE*	BN	12228	**VE**	E	*VE*	BN
12212	**VE**	E	*VE*	BN	12229	**EC**	E	*VE*	BN
12213	**VE**	E	*VE*	BN	12230	**VE**	E	*VE*	BN
12214	**EC**	E	*VE*	BN	12231	**EC**	E	*VE*	BN
12215	**VE**	E	*VE*	BN	12232	**EC**	E	*VE*	BN
12216	**VE**	E	*VE*	BN					

OPEN STANDARD (DISABLED)

Mark 4. Air conditioned. Rebuilt with new interior by Bombardier Wakefield 2003–05. –/68 2W 1TD. BT41 bogies. ETS 6X.

12331 was converted from Open Standard 12531.

12300–330. Lot No. 31048 Metro-Cammell 1989–91. 39.4 t.
12331. Lot No. 31049 Metro-Cammell 1989–92. 39.4 t.

12300	**VE**	E	*VE*	BN	12317	**VE**	E	*VE*	BN
12301	**VE**	E	*VE*	BN	12318	**VE**	E	*VE*	BN
12302	**EC**	E	*VE*	BN	12319	**VE**	E	*VE*	BN
12303	**VE**	E	*VE*	BN	12320	**VE**	E	*VE*	BN
12304	**VE**	E	*VE*	BN	12321	**VE**	E	*VE*	BN
12305	**EC**	E	*VE*	BN	12322	**VE**	E	*VE*	BN
12307	**VE**	E	*VE*	BN	12323	**VE**	E	*VE*	BN
12308	**VE**	E	*VE*	BN	12324	**VE**	E	*VE*	BN
12309	**VE**	E	*VE*	BN	12325	**VE**	E	*VE*	BN
12310	**VE**	E	*VE*	BN	12326	**VE**	E	*VE*	BN
12311	**VE**	E	*VE*	BN	12327	**VE**	E	*VE*	BN
12312	**EC**	E	*VE*	BN	12328	**VE**	E	*VE*	BN
12313	**VE**	E	*VE*	BN	12329	**EC**	E	*VE*	BN
12315	**VE**	E	*VE*	BN	12330	**VE**	E	*VE*	BN
12316	**VE**	E	*VE*	BN	12331	**VE**	E	*VE*	BN

OPEN STANDARD

Mark 4. Air conditioned. Rebuilt with new interior by Bombardier Wakefield 2003–05. –/76 1T. BT41 bogies. ETS 6X.

12405 is the second coach to carry that number. It was built from the bodyshell originally intended for 12221. The original 12405 is now 12232.

Lot No. 31049 Metro-Cammell 1989–92. 40.8 t.

12400	**VE**	E	*VE*	BN	12422	**VE**	E	*VE*	BN
12401	**VE**	E	*VE*	BN	12423	**EC**	E	*VE*	BN
12402	**EC**	E	*VE*	BN	12424	**VE**	E	*VE*	BN
12403	**VE**	E	*VE*	BN	12425	**VE**	E	*VE*	BN
12404	**VE**	E	*VE*	BN	12426	**VE**	E	*VE*	BN
12405	**EC**	E	*VE*	BN	12427	**VE**	E	*VE*	BN
12406	**VE**	E	*VE*	BN	12428	**VE**	E	*VE*	BN
12407	**VE**	E	*VE*	BN	12429	**VE**	E	*VE*	BN
12409	**VE**	E	*VE*	BN	12430	**VE**	E	*VE*	BN
12410	**EC**	E	*VE*	BN	12431	**VE**	E	*VE*	BN
12411	**EC**	E	*VE*	BN	12432	**VE**	E	*VE*	BN
12414	**VE**	E	*VE*	BN	12433	**VE**	E	*VE*	BN
12415	**VE**	E	*VE*	BN	12434	**VE**	E	*VE*	BN
12417	**VE**	E	*VE*	BN	12436	**VE**	E	*VE*	BN
12419	**EC**	E	*VE*	BN	12437	**VE**	E	*VE*	BN
12420	**VE**	E	*VE*	BN	12438	**VE**	E	*VE*	BN
12421	**VE**	E	*VE*	BN	12439	**VE**	E	*VE*	BN

12440	**VE**	E	*VE*	BN		12469	**VE**	E	*VE*	BN
12441	**VE**	E	*VE*	BN		12470	**VE**	E	*VE*	BN
12442	**VE**	E	*VE*	BN		12471	**VE**	E	*VE*	BN
12443	**EC**	E	*VE*	BN		12472	**VE**	E	*VE*	BN
12444	**VE**	E	*VE*	BN		12473	**VE**	E	*VE*	BN
12445	**VE**	E	*VE*	BN		12474	**VE**	E	*VE*	BN
12446	**VE**	E	*VE*	BN		12476	**VE**	E	*VE*	BN
12447	**VE**	E	*VE*	BN		12477	**VE**	E	*VE*	BN
12448	**EC**	E	*VE*	BN		12478	**VE**	E	*VE*	BN
12449	**VE**	E	*VE*	BN		12480	**VE**	E	*VE*	BN
12450	**EC**	E	*VE*	BN		12481	**VE**	E	*VE*	BN
12452	**VE**	E	*VE*	BN		12483	**VE**	E	*VE*	BN
12453	**VE**	E	*VE*	BN		12484	**VE**	E	*VE*	BN
12454	**VE**	E	*VE*	BN		12485	**VE**	E	*VE*	BN
12455	**VE**	E	*VE*	BN		12486	**VE**	E	*VE*	BN
12456	**VE**	E	*VE*	BN		12488	**EC**	E	*VE*	BN
12457	**VE**	E	*VE*	BN		12489	**EC**	E	*VE*	BN
12458	**VE**	E	*VE*	BN		12513	**VE**	E	*VE*	BN
12459	**VE**	E	*VE*	BN		12514	**VE**	E	*VE*	BN
12460	**VE**	E	*VE*	BN		12515	**VE**	E	*VE*	BN
12461	**VE**	E	*VE*	BN		12518	**VE**	E	*VE*	BN
12462	**VE**	E	*VE*	BN		12519	**VE**	E	*VE*	BN
12463	**VE**	E	*VE*	BN		12520	**VE**	E	*VE*	BN
12464	**VE**	E	*VE*	BN		12522	**VE**	E	*VE*	BN
12465	**VE**	E	*VE*	BN		12526	**EC**	E	*VE*	BN
12466	**VE**	E	*VE*	BN		12533	**VE**	E	*VE*	BN
12467	**VE**	E	*VE*	BN		12534	**VE**	E	*VE*	BN
12468	**VE**	E	*VE*	BN		12538	**VE**	E	*VE*	BN

OPEN STANDARD

Mark 3A. Air conditioned. Rebuilt for Chiltern Railways 2011–13 and fitted with sliding plug doors and toilets with retention tanks. Original InterCity 70 seating retained but mainly arranged around tables. –/72(6) or * –/69(4) 1T. BT10 bogies. ETS 6X.

12602–609/614–616/618/620. Lot No. 30877 Derby 1975–77. 36.2 t (* 37.1 t).
12601/613/617–619/621/623/625/627. Lot No. 30878 Derby 1975–76. 36.2 t (* 37.1 t).

12602	(12072)		**CM**	AV	*CR*	AL
12603	(12053)	*	**CM**	AV	*CR*	AL
12604	(12131)		**CM**	AV	*CR*	AL
12605	(11040)	*	**CM**	AV	*CR*	AL
12606	(12048)		**CM**	AV	*CR*	AL
12607	(12038)	*	**CM**	AV	*CR*	AL
12608	(12069)		**CM**	AV	*CR*	AL
12609	(12014)	*	**CM**	AV	*CR*	AL
12610	(12117)		**CM**	AV	*CR*	AL
12613	(11042, 12173)	*	**CM**	AV	*CR*	AL

12614–17018 51

12614 (12145)		**CM**	AV	*CR*	AL
12615 (12059)	*	**CM**	AV	*CR*	AL
12616 (12127)		**CM**	AV	*CR*	AL
12617 (11052, 12174)	*	**CM**	AV	*CR*	AL
12618 (11008, 12169)		**CM**	AV	*CR*	AL
12619 (11058, 12175)	*	**CM**	AV	*CR*	AL
12620 (12124)		**CM**	AV	*CR*	AL
12621 (11046)	*	**CM**	AV	*CR*	AL
12623 (11019)	*	**CM**	AV	*CR*	AL
12625 (11030)	*	**CM**	AV	*CR*	AL
12627 (11054)	*	**CM**	AV	*CR*	AL

CORRIDOR FIRST

Mark 1. Seven compartments. 42/– 2T. B4 bogies. ETS 3.

Lot No. 30381 Swindon 1959. 33 t.

13227	x	**CH**	LS		CL		13230	xk	**M** SP *SP*	BO
13229	xk	**M**	SP	*SP*	BO					

OPEN FIRST

Mark 1 converted from Corridor First in 2013–14. 42/– 2T. Commonwealth bogies. ETS 3.

Lot No. 30667 Swindon 1962. 35 t.

13320 x **M** WC *WC* CS ANNA

CORRIDOR FIRST

Mark 2A. Seven compartments. Pressure ventilated. 42/– 2T. B4 bogies. ETS 4.

Lot No. 30774 Derby 1968. 33 t.

13440 v **M** WC *WC* CS

CORRIDOR BRAKE FIRST

Mark 1. Four compartments. 24/– 1T. Commonwealth bogies. ETS 2.

Lot No. 30668 Swindon 1961. 36 t.

17018 (14018) v **CH** VT *VT* TM BOTAURUS

CORRIDOR BRAKE FIRST

Mark 2A. Four compartments. Pressure ventilated. 24/– 1T. B4 bogies. ETS 4.

17080/090 were numbered 35516/503 for a time when declassified.

17056/077. Lot No. 30775 Derby 1967–68. 32 t.
17080–102. Lot No. 30786 Derby 1968. 32 t.

52   17056–18756

17056 (14056)		**M**	RV		CL
17077 (14077)		**RV**	RV		BQ
17080 (14080)		**PC**	RF	*ST*	CS
17090 (14090)	v	**CH**	VT		TM
17102 (14102)		**M**	WC	*WC*	CS

COUCHETTE/GENERATOR COACH

Mark 2B. Formerly part of Royal Train. Converted from Corridor Brake First built 1969. Consists of luggage accommodation, guard's compartment, 350 kW diesel generator and staff sleeping accommodation. Pressure ventilated. B5 bogies. ETS 5X.

Lot No. 30888 Wolverton 1977. 46 t.

17105 (14105, 2905)	**RB**	RV	*RV*	EH

CORRIDOR BRAKE FIRST

Mark 2D. Four compartments. Air conditioned. Stones equipment. 24/- 1T. B4 Bogies. ETS 5.

Lot No. 30823 Derby 1971–72. 33.5 t.

17159 (14159)	d	**DS**	DR	*DR*	KM	
17167 (14167)		**VN**	BE	*NB*	CP	MOW COP

OPEN BRAKE UNCLASSIFIED

Mark 3B. Air conditioned. Fitted with hydraulic handbrake. Used by First Great Western as Sleeper "day coaches" that have been fitted with former HST First Class seats in a 2+1 layout and are effectively unclassified. 36/- 1T. BT10 bogies. pg. d. ETS 5X.

Lot No. 30990 Derby 1986. 35.81 t.

17173	**GW**	P	*GW*	PZ		17175	**GW**	P	*GW*	PZ
17174	**FD**	P	*GW*	PZ					

CORRIDOR STANDARD

Mark 1. –/48 2T. Eight Compartments. Commonwealth bogies. ETS 4.

Lot No. 30685 Derby 1961–62. 36 t.

18756 (25756)	x	**M**	WC	*WC*		CS

CORRIDOR BRAKE COMPOSITE

Mark 1. There are two variants depending upon whether the Standard Class compartments have armrests. Each vehicle has two First Class and three Standard Class compartments. 12/18 2T (* 12/24 2T). Commonwealth bogies. ETS 2.

21241–245. Lot No. 30669 Swindon 1961–62. 36 t.
21256. Lot No. 30731 Derby 1963. 37 t.
21266–272. Lot No. 30732 Derby 1964. 37 t.

21241	x	**M**	SP	*SP*	BO	21266	x*	**M**	WC	*WC*	CS
21245	x	**M**	RV	*RV*	EH	21269	*	**CC**	RV	*RV*	EH
21256	x	**M**	WC	*WC*	CS	21272	x*	**CH**	RV	*RV*	EH

CORRIDOR BRAKE STANDARD

Mark 1. Four compartments. –/24 1T. ETS 2.

35185. Lot No. 30427 Wolverton 1959. B4 bogies. 33 t.
35459. Lot No. 30721 Wolverton 1963. Commonwealth bogies. 37 t.

35185	x	**M**	SP	*SP*	BO
35459	x	**M**	WC	*WC*	CS

CORRIDOR BRAKE GENERATOR STANDARD

Mark 1. Four compartments. –/24 1T. Fitted with an ETS generator in the former luggage compartment.

Lot No. 30721 Wolverton 1963. Commonwealth bogies. 37 t.

35469 x **CC** RV *RV* EH

BRAKE/POWER KITCHEN

Mark 2C. Pressure ventilated. Converted from Corridor Brake First (declassified to Corridor Brake Standard) built 1970. Converted by West Coast Railway Company 2000–01. Consists of 60 kVA generator, guard's compartment and electric kitchen. B5 bogies. ETS 4.

Non-standard livery: British Racing Green with gold lining.

Lot No. 30796 Derby 1969–70. 32.5 t.

35511 (14130, 17130) **0** RF BU

KITCHEN CAR

Mark 1. Converted 1989/2006 from Kitchen Buffet Unclassified. Buffet and seating area replaced with additional kitchen and food preparation area. Fluorescent lighting. Commonwealth bogies. ETS 2X.

Lot No. 30628 Pressed Steel 1960–61. 39 t.

| 80041 | (1690) | x | **M** | RV | | EH |
| 80042 | (1646) | | **BG** | RV | *RV* | EH |

DRIVING BRAKE VAN (110 mph)

Mark 3B. Air conditioned. T4 bogies. dg. ETS 5X. Driving Brake Vans converted for use by Network Rail can be found in the Service Stock section of this book.

Non-standard livery: 82146 All over silver with DB logos.

Lot No. 31042 Derby 1988. 45.2 t.

82101	**V**	DR		BH	82123	**V**	AV		LM
82102	**1**	P	*GA*	NC	82126	**VT**	P	*GA*	NC
82103	**GA**	P	*GA*	NC	82127	**1**	P	*GA*	NC
82105	**GA**	P	*GA*	NC	82132	**GA**	P	*GA*	NC
82106	**V**	AV		LB	82133	**1**	P	*GA*	NC
82107	**GA**	P	*GA*	NC	82136	**GA**	P	*GA*	NC
82110	**V**	AV		LM	82137	**V**	AV		LM
82112	**GA**	P	*GA*	NC	82138	**V**	AV		LM
82113	**V**	AV		LM	82139	**GA**	P	*GA*	NC
82114	**GA**	P	*GA*	NC	82141	**V**	AV		LM
82115	**B**	NR		ZN	82143	**GA**	P	*GA*	NC
82116	**V**	AV		LM	82146	**0**	DB	*DB*	TO
82118	**GA**	P	*GA*	NC	82148	**V**	AV		LM
82120	**V**	AV		LM	82150	**V**	AV		LM
82121	**GA**	P	*GA*	NC	82152	**GA**	P	*GA*	NC
82122	**V**	AV		LM					

DRIVING BRAKE VAN (140 mph)

Mark 4. Air conditioned. Swiss-built (SIG) bogies. dg. ETS 6X.

Advertising livery: 82205 Flying Scotsman (purple).

Lot No. 31043 Metro-Cammell 1988. 43.5 t.

82200	**VE**	E	*VE*	BN	82216	**VE**	E	*VE*	BN
82201	**VE**	E	*VE*	BN	82217	**VE**	E	*VE*	BN
82202	**EC**	E	*VE*	BN	82218	**VE**	E	*VE*	BN
82203	**VE**	E	*VE*	BN	82219	**VE**	E	*VE*	BN
82204	**EC**	E	*VE*	BN	82220	**VE**	E	*VE*	BN
82205	**AL**	E	*VE*	BN	82222	**VE**	E	*VE*	BN
82206	**EC**	E	*VE*	BN	82223	**VE**	E	*VE*	BN
82207	**VE**	E	*VE*	BN	82224	**VE**	E	*VE*	BN
82208	**VE**	E	*VE*	BN	82225	**VE**	E	*VE*	BN
82209	**EC**	E	*VE*	BN	82226	**VE**	E	*VE*	BN
82210	**EC**	E	*VE*	BN	82227	**VE**	E	*VE*	BN
82211	**VE**	E	*VE*	BN	82228	**VE**	E	*VE*	BN
82212	**VE**	E	*VE*	BN	82229	**VE**	E	*VE*	BN
82213	**VE**	E	*VE*	BN	82230	**VE**	E	*VE*	BN
82214	**VE**	E	*VE*	BN	82231	**VE**	E	*VE*	BN
82215	**EC**	E	*VE*	BN					

82301–96175　　　　　　　　　　　　　　　　　　　　　　　　　　　　　　55

DRIVING BRAKE VAN (110 mph)

Mark 3B. Air conditioned. T4 bogies. dg. ETS 6X.

82301–305 originally converted 2008 for use by Wrexham & Shropshire. Now operated by Chiltern Railways. 82306–308 converted for Arriva Trains Wales 2011–12. 82309 converted for Chiltern Railways 2013.

g　Fitted with a diesel generator. 48.5 t.

Lot No. 31042 Derby 1988. 45.2 t.

82301	(82117)	g	**CM** AV	*CR*	AL
82302	(82151)	g	**CM** AV	*CR*	AL
82303	(82135)	g	**CM** AV	*CR*	AL
82304	(82130)	g	**CM** AV	*CR*	AL
82305	(82134)	g	**CM** AV	*CR*	AL
82306	(82144)		**AW** AV	*AW*	CF
82307	(82131)		**AW** AV	*AW*	CF
82308	(82108)		**AW** AV	*AW*	CF
82309	(82104)	g	**CM** AV	*CR*	AL

GANGWAYED BRAKE VAN (100 mph)

Mark 1. Short frame (57 ft). Load 10 t. Adapted 199? for use as Brake Luggage Van. Guard's compartment retained and former baggage area adapted for secure stowage of passengers' luggage. B4 bogies. 100 mph. ETS 1X.

Lot No. 30162 Pressed Steel 1956–57. 30.5 t.

92904　(80867, 99554)　　　**VN**　BE　*NB*　　　CP

HIGH SECURITY GENERAL UTILITY VAN

Mark 1. Modified with new floors, three roller shutter doors per side and the end doors removed. Commonwealth bogies. ETS 0X.

Lot No. 30616 Pressed Steel 1959–60. 32 t.

94225　(86849, 93849)　　　**M**　WC　*WC*　　　CS

GENERAL UTILITY VAN (100 mph)

Mark 1. Short frame. Load 14 t. Screw couplers. Adapted 2013/2010 for use as a water carrier with 3000 gallon capacity. ETS 0.

Non-standard livery: 96100 GWR Brown.

96100. Lot No. 30565 Pressed Steel 1959. 30 t. B5 bogies.
96175. Lot No. 30403 York/Glasgow 1958–60. 30 t. Commonwealth bogies.

96100　(86734, 93734)　x　**0**　VT　*VT*　　　TM
96175　(86628, 93628)　x　**M**　WC　*WC*　　　CS

KITCHEN CAR

Mark 1 converted from Corridor First in 2008 with staff accommodation. Commonwealth bogies. ETS 3.

Lot No. 30667 Swindon 1961. 35 t.

99316 (13321) x **M** WC *WC* CS

BUFFET STANDARD

Mark 1 converted from Open Standard in 2013 by the removal of two seating bays and fitting of a buffet. –/48 2T. Commonwealth bogies. ETS 4.

Lot No. 30646 Wolverton 1961. 36 t.

99318 (4912) x **M** WC *WC* CS

KITCHEN CAR

Mark 1 converted from Corridor Standard in 2011 with staff accommodation. Commonwealth bogies. ETS 3.

Lot No. 30685 Derby 1961–62. 34 t.

99712 (18893) x **M** WC *WC* CS

OPEN STANDARD

Mark 1 Corridor Standard rebuilt in 1997 as Open Standard using components from 4936. –/64 2T. Commonwealth bogies. ETS 4.

Lot No. 30685 Derby 1961–62. 36 t.

99722 (25806, 18806) x **M** WC *WC* CS

PLATFORM 5 MAIL ORDER

NARROW GAUGE STEAM LOCOMOTIVES
of Great Britain & Ireland

By Peter Nicholson

This completely new book from Platform 5 Publishing is the definitive guide to all narrow gauge steam locomotives known to exist in Great Britain and Ireland. Contains detailed lists of all locomotives including:

- Builder's Numbers
- Year of Manufacture
- Type & Class
- Gauge
- Current Location
- Running Numbers & Names

Further detailed notes describe each locomotive's origin and history. Also includes 64 colour illustrations, locomotive names index and a full list of heritage railways, museums and collections. 80 pages. **£14.95**

Also available:
**Miniature Railways of Great Britain & Ireland
(Published 2012) £14.95**

Please add postage: 10% UK, 20% Europe, 30% Rest of World.

Telephone, fax or send your order to the Platform 5 Mail Order Department. See inside back cover of this book for details.

NYMR REGISTERED CARRIAGES

These carriages are permitted to operate on the national railway network but may only be used to convey fare-paying passengers between Middlesbrough and Whitby on the Esk Valley branch as an extension of North Yorkshire Moors Railway services between Pickering and Grosmont. Only NYMR coaches currently registered for use on the national railway network are listed.

RESTAURANT FIRST

Mark 1. 24/–. Commonwealth bogies. Lot No. 30633 Swindon 1961. 42.5 t.

| 324 | x | **PC** | NY | *NY* | NY | | JOS de CRAU |

BUFFET STANDARD

Mark 1. –/44 2T. Commonwealth bogies.
Lot No. 30520 Wolverton 1960. 38 t.

| 1823 | v | **M** | NY | *NY* | NY |

OPEN STANDARD

Mark 1. –/64 2T (* –/60 2W 2T, † –/60 3W 1T). BR Mark 1 bogies.

3798/3801. Lot No. 30079 York 1953. 33 t.
3860/72. Lot No. 30080 York 1954. 33 t.
3948. Lot No. 30086 Eastleigh 1954–55. 33 t.
4198/4252. Lot No. 30172 York 1956. 33 t.
4286/90. Lot No. 30207 BRCW 1956. 33 t.
4455. Lot No. 30226 BRCW 1957. 33 t.

3798	v	**M**	NY	*NY*	NY		4198	v	**CC**	NY	*NY*	NY
3801	v	**CC**	NY	*NY*	NY		4252	v*	**CC**	NY	*NY*	NY
3860	v*	**M**	NY	*NY*	NY		4286	v	**CC**	NY	*NY*	NY
3872	vt	**BG**	NY	*NY*	NY		4290	v	**M**	NY	*NY*	NY
3948	v	**CC**	NY	*NY*	NY		4455	v	**CC**	NY	*NY*	NY

OPEN STANDARD

Mark 1. –/48 2T. BR Mark 1 bogies.
4786. Lot No. 30376 York 1957. 33 t.
4817. Lot No. 30473 BRCW 1959. 33 t.

| 4786 | v | **CH** | NY | *NY* | NY | | 4817 | v | **M** | NY | *NY* | NY |

OPEN STANDARD

Mark 1. Later vehicles built with Commonwealth bogies. –/64 2T.
Lot No. 30690 Wolverton 1961–62. Aluminium window frames. 37 t.

| 4990 | v | **M** | NY | *NY* | NY | | 5029 | v | **CH** | NY | *NY* | NY |
| 5000 | v | **M** | NY | *NY* | NY |

NYMR REGISTERED CARRIAGES

OPEN BRAKE STANDARD

Mark 1. –/39 1T. BR Mark 1 bogies.
Lot No. 30170 Doncaster 1956. 34 t.

| 9225 | v | **M** | NY | *NY* | NY | | 9274 | v | **M** | NY | *NY* | NY |
| 9267 | v | **BG** | NY | *NY* | NY | | | | | | | |

CORRIDOR COMPOSITE

Mark 1. 24/18 1T. BR Mark 1 bogies.
15745. Lot No. 30179 Metro Cammell 1956. 36 t.
16156. Lot No. 30665 Derby 1961. 36 t.

| 15745 | v | **M** | NY | *NY* | NY | | 16156 | v | **CC** | NY | *NY* | NY |

CORRIDOR BRAKE COMPOSITE

Mark 1. Two First Class and three Standard Class compartments. 12/18 2T. BR Mark 1 bogies.
Lot No. 30185 Metro Cammell 1956. 36 t.

21100 v **CC** NY *NY* NY

CORRIDOR BRAKE STANDARD

Mark 1. –/24 1T. BR Mark 1 bogies.
Lot No. 30233 Gloucester 1957. 35 t.

35089 v **CC** NY *NY* NY

PULLMAN BRAKE THIRD

Built 1928 by Metropolitan Carriage & Wagon Company. –/30. Gresley bogies. 37.5 t.

232 v **PC** NY *NY* NY CAR No. 79

PULLMAN KITCHEN FIRST

Built by Metro-Cammell 1960–61 for ECML services. 20/– 2T. Commonwealth bogies. 41.2 t.

318 x **PC** NY *NY* NY ROBIN

PULLMAN PARLOUR FIRST

Built by Metro-Cammell 1960–61 for ECML services. 29/– 2T. Commonwealth bogies. 38.5 t.

328 x **PC** NY *NY* NY OPAL

2. HIGH SPEED TRAIN TRAILER CARS

HSTs consist of a number of trailer cars (usually between six and nine) with a power car at each end. All trailers are classified Mark 3 and have BT10 bogies with disc brakes and central door locking. Heating is by a 415 V three-phase supply and vehicles have air conditioning. Maximum speed is 125 mph.

The trailer cars have one standard bodyshell for both First and Standard Class, thus facilitating easy conversion from one class to the other. As built all cars had facing seating around tables with Standard Class carriages having nine bays of seats per side which did not line up with the eight windows per side. This created a new unwelcome trend in British rolling stock of seats not lining up with windows.

All vehicles underwent a mid-life refurbishment in the 1980s with Standard Class seating layouts revised to incorporate unidirectional seating in addition to facing, in a somewhat higgledy-piggledy layout where seats did not line up either side of the aisle.

A further refurbishment programme was completed in November 2000, with each Train Operating Company having a different scheme as follows:

Great Western Trains (later First Great Western). Green seat covers and extra partitions between seat bays.

Great North Eastern Railway. New ceiling lighting panels and brown seat covers. First Class vehicles had table lamps and imitation walnut plastic end panels.

Virgin CrossCountry. Green seat covers. Standard Class vehicles had four seats in the centre of each carriage replaced with a luggage stack. All have now passed to other operators.

Midland Mainline. Grey seat covers, redesigned seat squabs, side carpeting and two seats in the centre of each Standard Class carriage and one in First Class carriages replaced with a luggage stack.

Since then the remaining three operators of HSTs embarked on separate, and very different, refurbishment projects:

Midland Mainline was first to refurbish its vehicles a second time during 2003-04. This involved fitting new fluorescent and halogen ceiling lighting, although the original seats were retained in First and Standard Class, but with blue upholstery.

London St Pancras-Sheffield/Leeds and Nottingham services are now operated by **East Midlands Trains** and in late 2009 this operator embarked on another, less radical, refurbishment which included retention of the original seats but with red upholstery in Standard Class and blue in First Class. This programme was completed in 2010.

First Great Western (now **Great Western Railway**) started a major rebuild of its HST sets in late 2006, with the programme completed in 2008. With an increased fleet of 54 sets (since reduced to 53) the new interiors feature new lighting and seating throughout. First Class seats have leather upholstery, and are made by Primarius UK. Standard Class seats are of high-back design by Grammer. A number of sets operate without a full buffet or kitchen car, instead using one of 19 TS vehicles converted to include a "mini buffet" counter for use on shorter distance services. During 2012 15 402xx or 407xx buffet vehicles were converted to Trailer Standards to make the rakes formed as 7-cars up to 8-cars. During 2014–15 further changes to the FGW sets include the conversion of one First Class carriage from each set to Standard Class.

GNER modernised its buffet cars with new corner bars in 2004 and at the same time each HST set was made up to 9-cars with an extra Standard Class vehicle added with a disabled person's toilet.

At the end of 2006 **GNER** embarked on a major rebuild of its sets, with the work being carried out at Wabtec, Doncaster. All vehicles will have similar interiors to the Mark 4 "Mallard" fleet, with new Primarius seats throughout. The refurbishment of the 13 sets was completed by **National Express East Coast** in late 2009, these trains are now operated by **Virgin Trains East Coast**.

Ten sets ex-Virgin CrossCountry, and some spare vehicles, were temporarily allocated to Midland Mainline for the interim service to Manchester during 2003–04 and had a facelift. Buffet cars were converted from TSB to TFKB and renumbered in the 408xx series. Most of these vehicles are now in use with Great Western Railway.

Open access operator **Grand Central** started operation in December 2007 with a new service from Sunderland to London King's Cross. This operator has three sets mostly using stock converted from loco-hauled Mark 3s. The seats in Standard Class have First Class spacing and in most vehicles are all facing. Increases in the number of passengers on its services has seen rakes lengthened from 5-cars to 6-cars – they can run as 7-cars at busy times.

CrossCountry reintroduced HSTs to the Cross-Country network from 2008. Five sets were refurbished at Wabtec, Doncaster principally for use on the Plymouth–Edinburgh route. Three of these sets use stock mostly converted from loco-hauled Mark 3s and two are sets ex-Midland Mainline. The interiors are similar to refurbished East Coast sets, although the seating layout is different and one toilet per carriage has been removed in favour of a luggage stack.

Operator Codes

Operator codes are shown in the heading before each set of vehicles. The first letter is always "T" for HST carriages, denoting a Trailer vehicle. The second letter denotes the passenger accommodation in that vehicle, for example "F" for First. "GS" denotes Guards accommodation and Standard Class seating. This is followed by catering provision, with "B" for buffet, and "K" for a kitchen and buffet:

TCK	Trailer Composite Kitchen		TFKB	Trailer Kitchen Buffet First
TF	Trailer First		TSB	Trailer Buffet Standard
TFB	Trailer Buffet First		TS	Trailer Standard
TGS	Trailer Guard's Standard			

TRAILER STANDARD BUFFET　　　　　　　　　　　TSB

19 vehicles converted at Laira 2009-10 from HST TSs for First Great Western. Refurbished with Grammer seating.

40101–119. For Lot No. details see TS. –/70 1T. 35.5 t.

40101 (42170)	**FD**	P	*GW*	LA
40102 (42223)	**FD**	P	*GW*	LA
40103 (42316)	**FD**	P	*GW*	LA
40104 (42254)	**FD**	P	*GW*	LA
40105 (42084)	**FD**	P	*GW*	LA
40106 (42162)	**FD**	P	*GW*	LA
40107 (42334)	**FD**	P	*GW*	LA
40108 (42314)	**FD**	P	*GW*	LA
40109 (42262)	**FD**	P	*GW*	LA
40110 (42187)	**FD**	P	*GW*	LA
40111 (42248)	**FD**	P	*GW*	LA
40112 (42336)	**FD**	P	*GW*	LA
40113 (42309)	**FD**	P	*GW*	LA
40114 (42086)	**FD**	P	*GW*	LA
40115 (42320)	**FD**	P	*GW*	LA
40116 (42147)	**FD**	P	*GW*	LA
40117 (42249)	**FD**	P	*GW*	LA
40118 (42338)	**FD**	P	*GW*	LA
40119 (42090)	**FD**	P	*GW*	LA

TRAILER BUFFET FIRST　　　　　　　　　　　　TFB

Converted from TSB by fitting First Class seats. Renumbered from 404xx series by subtracting 200. All refurbished by First Great Western and fitted with Primarius leather seating. 23/–.

40204–221. Lot No. 30883 Derby 1976–77. 36.12 t.
40231. Lot No. 30899 Derby 1978–79. 36.12 t.

40204	**FD**	A	*GW*	LA		40210	**FD**	A	*GW*	LA
40205	**FD**	A	*GW*	LA		40221	**FD**	A	*GW*	LA
40207	**FD**	A	*GW*	LA		40231	**FD**	A	*GW*	LA

TRAILER BUFFET STANDARD　　　　　　　　　TSB

Renumbered from 400xx series by adding 400. –/33 1W.

40433 was numbered 40233 for a time when fitted with 23 First Class seats.

40402–426. Lot No. 30883 Derby 1976–77. 36.12 t.
40433. Lot No. 30899 Derby 1978–79. 36.12 t.

40402	**V**	AV		LM		40426	**GC**	A	*GC*	HT
40419	**V**	AV		LM		40433	**GC**	A	*GC*	HT
40424	**GC**	A	*GC*	HT						

TRAILER KITCHEN BUFFET FIRST　　　　　　　　　　　TFKB

These vehicles have larger kitchens than the 402xx and 404xx series vehicles, and are used in trains where a full meal service is required. They were renumbered from the 403xx series (in which the seats were unclassified) by adding 400 to the previous number. 17/–.

* Refurbished Great Western Railway vehicles. Primarius leather seating.
m Refurbished Virgin Trains East Coast vehicles with Primarius seating.

40700–721. Lot No. 30921 Derby 1978–79. 38.16 t.
40722–735. Lot No. 30940 Derby 1979–80. 38.16 t.
40737–753. Lot No. 30948 Derby 1980–81. 38.16 t.
40754–757. Lot No. 30966 Derby 1982. 38.16 t.

40700		**ST**	P	*EM*	NL	40732	**EC**	P	*VE*	EC
40701	m	**VE**	P	*VE*	EC	40733 *	**FD**	A	*GW*	LA
40702	m	**NX**	P	*VE*	EC	40734 *	**FD**	A	*GW*	LA
40703	*	**FD**	A	*GW*	LA	40735 m	**VE**	A	*VE*	EC
40704	m	**VE**	A	*VE*	EC	40737 m	**VE**	A	*VE*	EC
40705	m	**VE**	A	*VE*	EC	40739 *	**FD**	A	*GW*	LA
40706	m	**VE**	A	*VE*	EC	40740 m	**VE**	A	*VE*	EC
40707	*	**FD**	A	*GW*	LA	40741	**ST**	P	*EM*	NL
40708	m	**VE**	P	*VE*	EC	40742 m	**VE**	A	*VE*	EC
40710	*	**FD**	A	*GW*	LA	40743 *	**FD**	A	*GW*	LA
40711	m	**VE**	A	*VE*	EC	40746	**ST**	P	*EM*	NL
40713	*	**FD**	A	*GW*	LA	40748 m	**VE**	A	*VE*	EC
40715	*	**GW**	A	*GW*	LA	40749	**ST**	P	*EM*	NL
40716	*	**FD**	A	*GW*	LA	40750 m	**NX**	A	*VE*	EC
40718	*	**FD**	A	*GW*	LA	40751	**VE**	P	*VE*	NL
40720	m	**VE**	A	*VE*	EC	40752 *	**FD**	A	*GW*	LA
40721	*	**FD**	A	*GW*	LA	40753	**ST**	P	*EM*	NL
40722	*	**FD**	A	*GW*	LA	40754	**ST**	P	*EM*	NL
40727	*	**FD**	A	*GW*	LA	40755 *	**FD**	A	*GW*	LA
40728		**ST**	P	*EM*	NL	40756	**ST**	P	*EM*	NL
40730		**ST**	P	*EM*	NL	40757 *	**FD**	A	*GW*	LA

TRAILER KITCHEN BUFFET FIRST　　　　　　　　　　　TFKB

These vehicles have been converted from TSBs in the 404xx series to be similar to the 407xx series vehicles. 17/–.

40802/804/811 were numbered 40212/232/211 for a time when fitted with 23 First Class seats.

* Refurbished Great Western Railway vehicles. Primarius leather seating.
m Refurbished Virgin Trains East Coast vehicle with Primarius seating.

40801–803/805/808/809/811. Lot No. 30883 Derby 1976–77. 38.16 t.
40804/806/807/810. Lot No. 30899 Derby 1978–79. 38.16 t.

40801	(40027, 40427)	*	**FD**	P	*GW*	LA
40802	(40012, 40412)	*	**FD**	P	*GW*	LA

40803	(40018, 40418)	*	**FD** P	*GW*	LA
40804	(40032, 40432)	*	**FD** P	*GW*	LA
40805	(40020, 40420)	m	**NX** P		NL
40806	(40029, 40429)	*	**FD** P	*GW*	LA
40807	(40035, 40435)	*	**FD** P	*GW*	LA
40808	(40015, 40415)	*	**FD** P	*GW*	LA
40809	(40014, 40414)	*	**FD** P	*GW*	LA
40810	(40030, 40430)	*	**FD** P	*GW*	LA
40811	(40011, 40411)	*	**FD** P	*GW*	LA

TRAILER BUFFET FIRST — TFB

Vehicles owned by First Group. Converted from TSB by First Great Western. Refurbished with Primarius leather seating. 23/–.

40900/902/904. Lot No. 30883 Derby 1976–77. 36.12 t.
40901/903. Lot No. 30899 Derby 1978–79. 36.12 t.

40900	(40022, 40422)	**FD** FG	*GW*	LA
40901	(40036, 40436)	**FD** FG	*GW*	LA
40902	(40023, 40423)	**FD** FG	*GW*	LA
40903	(40037, 40437)	**FD** FG	*GW*	LA
40904	(40001, 40401)	**FD** FG	*GW*	LA

TRAILER FIRST — TF

As built and m 48/– 2T (m† 48/– 1T – one toilet removed for trolley space).
* Refurbished Great Western Railway vehicles. Primarius leather seating.
c Refurbished CrossCountry vehicles with Primarius seating and 2 tip-up seats. One toilet removed. 40/– 1TD 1W.
m Refurbished Virgin Trains East Coast vehicles with Primarius seating.
s Fitted with centre luggage stack, disabled toilet and wheelchair space. 46/– 1T 1TD 1W.
w Wheelchair space. 47/– 2T 1W. (41154 47/– 1TD 1T 1W).
x Toilet removed for trolley space (FGW). 48/– 1T.

41004–056. Lot No. 30881 Derby 1976–77. 33.66 t.
41057–120. Lot No. 30896 Derby 1977–78. 33.66 t.
41122–148. Lot No. 30938 Derby 1979–80. 33.66 t.
41149–166. Lot No. 30947 Derby 1980. 33.66 t.
41167–169. Lot No. 30963 Derby 1982. 33.66 t.
41170. Lot No. 30967 Derby 1982. Former prototype vehicle. 33.66 t.
41176. Lot No. 30897 Derby 1977. 33.66 t.
41180. Lot No. 30884 Derby 1976–77. 33.66 t.
41181–184/189. Lot No. 30939 Derby 1979–80. 33.66 t.
41185–187/191. Lot No. 30969 Derby 1982. 33.66 t.
41190. Lot No. 30882 Derby 1976–77. 33.66 t.
41192. Lot No. 30897 Derby 1977–79. 33.60 t.
41193–195/201–206. Lot No. 30878 Derby 1975–76. 34.3 t. Converted from Mark 3A Open First.

41004	*x	**FD**	A	*GW*	LA	41095	mw	**VE**	P	*VE*	EC
41006	*w	**FD**	A	*GW*	LA	41097	mt	**NX**	A	*VE*	EC
41008	*w	**FD**	A	*GW*	LA	41098	mw	**NX**	A	*VE*	EC
41010	*w	**FD**	A	*GW*	LA	41099	mt	**VE**	A	*VE*	EC
41012	*w	**FD**	A	*GW*	LA	41100	mw	**VE**	A	*VE*	EC
41016	*w	**FD**	A	*GW*	LA	41102	*w	**FD**	A	*GW*	LA
41018	*w	**FD**	A	*GW*	LA	41103	*x	**FD**	A	*GW*	LA
41020	*w	**FD**	A	*GW*	LA	41104	*w	**FD**	A	*GW*	LA
41022	*w	**FD**	A	*GW*	LA	41106	*w	**FD**	A	*GW*	LA
41024	*w	**FD**	A	*GW*	LA	41108	*w	**FD**	P	*GW*	LA
41026	c	**XC**	A	*XC*	EC	41110	*w	**FD**	A	*GW*	LA
41028	*w	**FD**	A	*GW*	LA	41111		**ST**	P	*EM*	NL
41030	*w	**FD**	A	*GW*	LA	41112		**VE**	P	*VE*	NL
41032	*w	**FD**	A	*GW*	LA	41113	s	**ST**	P	*EM*	NL
41034	*w	**FD**	A	*GW*	LA	41115	mt	**NX**	P	*VE*	EC
41035	c	**XC**	A	*XC*	EC	41116	*x	**FD**	A	*GW*	LA
41038	*w	**FD**	A	*GW*	LA	41117		**ST**	P	*EM*	NL
41039	mt	**VE**	A	*VE*	EC	41118	mw	**VE**	A	*VE*	EC
41040	mw	**VE**	A	*VE*	EC	41120	mt	**VE**	A	*VE*	EC
41041	s	**ST**	P	*EM*	NL	41122	*w	**FD**	A	*GW*	LA
41044	mw	**VE**	A	*VE*	EC	41124	*w	**FD**	A	*GW*	LA
41046	s	**ST**	P	*EM*	NL	41126	*w	**FD**	A	*GW*	LA
41052	*w	**FD**	A	*GW*	LA	41128	*w	**FD**	A	*GW*	LA
41056	*w	**FD**	A	*GW*	LA	41130	*w	**FD**	A	*GW*	LA
41057		**ST**	P	*EM*	NL	41132	*w	**FD**	A	*GW*	LA
41059	*w	**FD**	FG	*GW*	LA	41134	*w	**FD**	A	*GW*	LA
41061	w	**ST**	P	*EM*	NL	41135	*w	**FD**	A	*GW*	LA
41062		**EC**	P	*VE*	EC	41136	*w	**FD**	A	*GW*	LA
41063		**ST**	P	*EM*	NL	41137	*x	**FD**	A	*GW*	LA
41064	s	**ST**	P	*EM*	NL	41138	*w	**FD**	A	*GW*	LA
41066	mt	**VE**	A	*VE*	EC	41140	*w	**FD**	A	*GW*	LA
41067	s	**ST**	P	*EM*	NL	41142	*w	**FD**	A	*GW*	LA
41068	s	**VE**	P	*VE*	NL	41144	*w	**FD**	A	*GW*	LA
41069	s	**ST**	P	*EM*	NL	41146	*w	**GW**	A	*GW*	LA
41070	s	**ST**	P	*EM*	NL	41149	*w	**FD**	P	*GW*	LA
41071		**ST**	P	*EM*	NL	41150	mw	**VE**	A	*VE*	EC
41072	s	**ST**	P	*EM*	NL	41151	mt	**VE**	A	*VE*	EC
41075	s	**ST**	P	*EM*	NL	41152	mw	**VE**	A	*VE*	EC
41076	s	**ST**	P	*EM*	NL	41154	w	**EC**	P	*VE*	EC
41077		**ST**	P	*EM*	NL	41156		**ST**	P	*EM*	NL
41079		**ST**	P	*EM*	NL	41158	*w	**FD**	A	*GW*	LA
41083	mw	**VE**	P	*VE*	EC	41159	mt	**VE**	P	*VE*	EC
41084	s	**ST**	P	*EM*	NL	41160	*w	**FD**	FG	*GW*	LA
41087	mt	**VE**	A	*VE*	EC	41161	*w	**FD**	P	*GW*	LA
41088	mw	**VE**	A	*VE*	EC	41162	*w	**FD**	FG	*GW*	LA
41089	*w	**FD**	A	*GW*	LA	41164	mw	**VE**	A	*VE*	EC
41090	mw	**VE**	A	*VE*	EC	41165	mw	**NX**	P	*VE*	EC
41091	mt	**VE**	A	*VE*	EC	41166	*w	**FD**	FG	*GW*	LA
41092	mw	**VE**	A	*VE*	EC	41167	*w	**FD**	FG	*GW*	LA
41094	*w	**FD**	A	*GW*	LA	41169	*w	**FD**	P	*GW*	LA

41170	(41001)	mt	**VE**	A *VE*	EC
41176	(42142, 42352)	*w	**FD**	P *GW*	LA
41180	(40511)	*w	**FD**	A *GW*	LA
41182	(42278)	*w	**FD**	P *GW*	LA
41183	(42274)	*w	**FD**	P *GW*	LA
41185	(42313)	mt	**VE**	P *VE*	EC
41186	(42312)	*w	**FD**	P *GW*	LA
41187	(42311)	*w	**FD**	P *GW*	LA
41189	(42298)	*w	**FD**	P *GW*	LA
41190	(42088)	mt	**VE**	P *VE*	EC
41192	(42246)	*w	**FD**	P *GW*	LA

The following carriages have been converted from loco-hauled Mark 3 vehicles for CrossCountry or Grand Central.

41193	(11060)	c	**XC**	P *XC*	EC
41194	(11016)	c	**XC**	P *XC*	EC
41195	(11020)	c	**XC**	P *XC*	EC
41201	(11045)		**GC**	A *GC*	HT
41202	(11017)		**GC**	A *GC*	HT
41203	(11038)		**GC**	A *GC*	HT
41204	(11023)		**GC**	A *GC*	HT
41205	(11036)		**GC**	A *GC*	HT
41206	(11055)		**GC**	A *GC*	HT

TRAILER STANDARD — TS

42158 was numbered 41177 for a time when fitted with First Class seats.
42310 was numbered 41188 for a time when fitted with First Class seats.

Standard seating and m –/76 2T.
* Refurbished Great Western Railway vehicles. Grammer seating. –/80 2T (unless h – high density).
c Refurbished CrossCountry vehicles with Primarius seating. One toilet removed. –/82 1T.
§c Refurbished CrossCountry vehicles with Primarius seating and 2 tip-up seats. One toilet removed. –/66 1TD 2W. 42379/380 are –/71 1TD 1T 2W.
d Great Western Railway vehicles with disabled persons toilet and 5, 6 or 7 tip-up seats. –/68 1T 1TD 2W.
h "High density" Great Western Railway vehicles. –/84 2T.
k "High density" Great Western Railway refurbished vehicle with disabled persons toilet and 5, 6 or 7 tip-up seats. –/72 1T 1TD 2W.
m Refurbished Virgin Trains East Coast vehicles with Primarius seating.

42003–42047

u Centre luggage stack (EMT) –/74 2T.
w Centre luggage stack and wheelchair space (EMT) –72 2T 1W.
† Disabled persons toilet (VTEC) –/62 1T 1TD 1W.

42003–089/362. Lot No. 30882 Derby 1976–77. 33.6 t.
42091–250. Lot No. 30897 Derby 1977–79. 33.6 t.
42251–305. Lot No. 30939 Derby 1979–80. 33.6 t.
42306–322. Lot No. 30969 Derby 1982. 33.6 t.
42323–341. Lot No. 30983 Derby 1984–85. 33.6 t.
42342/360. Lot No. 30949 Derby 1982. 33.47 t. Converted from TGS.
42343/345. Lot No. 30970 Derby 1982. 33.47 t. Converted from TGS.
42344/361. Lot No. 30964 Derby 1982. 33.47 t. Converted from TGS.
42346/347/350/351/379/380/551–564. Lot No. 30881 Derby 1976–77. 33.66 t. Converted from TF.
42348/349/363–365/381/565–570. Lot No. 30896 Derby 1977–78. 33.66 t. Converted from TF.
42354. Lot No. 30897 Derby 1977. Was TF from 1983 to 1992. 33.66 t.
42353/355–357. Lot No. 30967 Derby 1982. Ex-prototype vehicles. 33.66 t.
42366–378/382/383/401–409. Lot No. 30877 Derby 1975–77. 34.3 t. Converted from Mark 3A Open Standard.
42384. Lot No. 30896 Derby 1977–78. 33.66 t. Converted from TF.
42385/580–583. Lot No. 30947 Derby 1980. 33.66 t. Converted from TF.
42501/509/513/515/517. Lot No. 30948 Derby 1980–81. 34.8 t. Converted from TFKB.
42502/506/508/514/516. Lot No. 30940 Derby 1979–80. 34.8 t. Converted from TFKB.
42503/504/510/511. Lot No. 30921 Derby 1978–79. 34.8 t. Converted from TFKB.
42505/507/512/518/519. Lot No. 30883 Derby 1976–77. 34.8 t. Converted from TSB.
42520. Lot No. 30899 Derby 1978–79. 34.8 t. Converted from TSB.
42571–579. Lot No. 30938 Derby 1979–80. 33.66 t. Converted from TF.

42003	*h	**FD**	A	*GW*	LA	42028	*h	**FD**	A	*GW*	LA
42004	*d	**FD**	A	*GW*	LA	42029	*h	**FD**	A	*GW*	LA
42005	*h	**FD**	A	*GW*	LA	42030	*k	**FD**	A	*GW*	LA
42006	*h	**FD**	A	*GW*	LA	42031	*h	**FD**	A	*GW*	LA
42007	*d	**FD**	A	*GW*	LA	42032	*h	**FD**	A	*GW*	LA
42008	*k	**FD**	A	*GW*	LA	42033	*	**FD**	A	*GW*	LA
42009	*h	**FD**	A	*GW*	LA	42034	*	**FD**	A	*GW*	LA
42010	*h	**FD**	A	*GW*	LA	42035	*	**FD**	A	*GW*	LA
42012	*k	**FD**	A	*GW*	LA	42036	c	**XC**	A	*XC*	EC
42013	*h	**FD**	A	*GW*	LA	42037	c	**XC**	A	*XC*	EC
42014	*h	**FD**	A	*GW*	LA	42038	c	**XC**	A	*XC*	EC
42015	*k	**FD**	A	*GW*	LA	42039	*h	**FD**	A	*GW*	LA
42016	*h	**FD**	A	*GW*	LA	42040	*h	**FD**	A	*GW*	LA
42019	*	**FD**	A	*GW*	LA	42041	*h	**FD**	A	*GW*	LA
42021	*k	**FD**	A	*GW*	LA	42042	*h	**FD**	A	*GW*	LA
42023	*h	**FD**	A	*GW*	LA	42043	*h	**FD**	A	*GW*	LA
42024	*k	**FD**	A	*GW*	LA	42044	*h	**FD**	A	*GW*	LA
42025	*h	**FD**	A	*GW*	LA	42045	*h	**FD**	A	*GW*	LA
42026	*h	**FD**	A	*GW*	LA	42046	*	**FD**	A	*GW*	LA
42027	*h	**FD**	A	*GW*	LA	42047	*	**FD**	A	*GW*	LA

42048	*h	**FD**	A	*GW*	LA	42104	m	**VE**	A	*VE*	EC
42049	*h	**FD**	A	*GW*	LA	42105	*k	**FD**	FG	*GW*	LA
42050	*h	**FD**	A	*GW*	LA	42106	m	**VE**	A	*VE*	EC
42051	c	**XC**	A	*XC*	EC	42107	*	**FD**	A	*GW*	LA
42052	c	**XC**	A	*XC*	EC	42108	*h	**FD**	FG	*GW*	LA
42053	c	**XC**	A	*XC*	EC	42109	m	**NX**	P	*VE*	EC
42054	*	**FD**	A	*GW*	LA	42110	m	**NX**	P	*VE*	EC
42055	*	**FD**	A	*GW*	LA	42111	u	**ST**	P	*EM*	NL
42056	*	**FD**	A	*GW*	LA	42112	u	**ST**	P	*EM*	NL
42057	m	**VE**	A	*VE*	EC	42113	u	**ST**	P	*EM*	NL
42058	m	**VE**	A	*VE*	EC	42115	*h	**FD**	P	*GW*	LA
42059	m	**VE**	A	*VE*	EC	42116	m†	**VE**	A	*VE*	EC
42060	*h	**FD**	A	*GW*	LA	42117	m	**NX**	P	*VE*	EC
42061	*h	**FD**	A	*GW*	LA	42118	*h	**FD**	A	*GW*	LA
42062	*k	**FD**	A	*GW*	LA	42119	u	**ST**	P	*EM*	NL
42063	m	**VE**	A	*VE*	EC	42120	u	**ST**	P	*EM*	NL
42064	m	**VE**	A	*VE*	EC	42121	u	**ST**	P	*EM*	NL
42065	m	**VE**	A	*VE*	EC	42122	m	**VE**	A	*VE*	EC
42066	*k	**FD**	A	*GW*	LA	42123		**EC**	P	*VE*	EC
42067	*h	**FD**	A	*GW*	LA	42124	u	**ST**	P	*EM*	NL
42068	*h	**FD**	A	*GW*	LA	42125		**EC**	P	*VE*	EC
42069	*k	**FD**	A	*GW*	LA	42126	*h	**FD**	A	*GW*	LA
42070	*h	**FD**	A	*GW*	LA	42127	m†	**VE**	A	*VE*	EC
42071	*h	**FD**	A	*GW*	LA	42128	m†	**VE**	A	*VE*	EC
42072	*	**FD**	A	*GW*	LA	42129	*	**FD**	A	*GW*	LA
42073	*h	**FD**	A	*GW*	LA	42130	m	**VE**	P	*VE*	EC
42074	*h	**FD**	A	*GW*	LA	42131	u	**ST**	P	*EM*	NL
42075	*	**FD**	A	*GW*	LA	42132	w	**ST**	P	*EM*	NL
42076	*	**FD**	A	*GW*	LA	42133	u	**ST**	P	*EM*	NL
42077	*	**FD**	A	*GW*	LA	42134	m	**VE**	A	*VE*	EC
42078	*	**FD**	A	*GW*	LA	42135	u	**ST**	P	*EM*	NL
42079	*h	**FD**	A	*GW*	LA	42136	u	**ST**	P	*EM*	NL
42080	*h	**FD**	A	*GW*	LA	42137	u	**ST**	P	*EM*	NL
42081	*k	**FD**	A	*GW*	LA	42138	*k	**FD**	A	*GW*	LA
42083	*h	**FD**	A	*GW*	LA	42139	u	**ST**	P	*EM*	NL
42085	*h	**FD**	P	*GW*	LA	42140	u	**ST**	P	*EM*	NL
42087	*h	**FD**	P	*GW*	LA	42141	u	**ST**	P	*EM*	NL
42089	*h	**FD**	A	*GW*	LA	42143	*	**FD**	A	*GW*	LA
42091	m†	**VE**	A		*VE*	EC	42144	*	**FD**	A	*GW*	LA
42092	*k	**FD**	FG	*GW*	LA	42145	*	**FD**	A	*GW*	LA
42093	*h	**FD**	FG	*GW*	LA	42146	m	**VE**	A	*VE*	EC
42094	*h	**FD**	FG	*GW*	LA	42148	u	**ST**	P	*EM*	NL
42095	*	**FD**	FG	*GW*	LA	42149	u	**ST**	P	*EM*	NL
42096	*h	**FD**	A	*GW*	LA	42150	m	**VE**	A	*VE*	EC
42097	c	**XC**	A	*XC*	EC	42151	w	**ST**	P	*EM*	NL
42098	*h	**FD**	A	*GW*	LA	42152	u	**ST**	P	*EM*	NL
42099	*h	**FD**	A	*GW*	LA	42153	u	**ST**	P	*EM*	NL
42100	u	**ST**	P	*EM*	NL	42154	m	**VE**	A	*VE*	EC
42101	*h	**FD**	P	*GW*	LA	42155	w	**ST**	P	*EM*	NL
42102	*h	**FD**	P	*GW*	LA	42156	u	**ST**	P	*EM*	NL
42103	*k	**FD**	FG	*GW*	LA	42157	u	**ST**	P	*EM*	NL

42158	m	**NX**	A	*VE*	EC	42212	*h	**FD**	A	*GW*	LA
42159	mt	**NX**	P	*VE*	EC	42213	*h	**FD**	A	*GW*	LA
42160	m	**NX**	P	*VE*	EC	42214	*h	**FD**	A	*GW*	LA
42161	mt	**VE**	A	*VE*	EC	42215	m	**VE**	A	*VE*	EC
42163	m	**VE**	P	*VE*	EC	42216	*h	**FD**	A	*GW*	LA
42164		**ST**	P	*EM*	NL	42217	*k	**FD**	P	*GW*	LA
42165		**ST**	P	*EM*	NL	42218	*k	**FD**	P	*GW*	LA
42166	*h	**FD**	P	*GW*	LA	42219	m	**VE**	A	*VE*	EC
42167	*h	**FB**	FG	*GW*	LA	42220	w	**ST**	P	*EM*	NL
42168	*h	**FB**	FG	*GW*	LA	42221	*h	**FD**	A	*GW*	LA
42169	*h	**FB**	FG	*GW*	LA	42222	*h	**FD**	P	*GW*	LA
42171	m	**VE**	A	*VE*	EC	42224	*k	**FD**	P	*GW*	LA
42172	m	**VE**	A	*VE*	EC	42225	u	**VE**	P	*VE*	NL
42173	*k	**FD**	P	*GW*	LA	42226	m	**VE**	A	*VE*	EC
42174	*k	**FD**	P	*GW*	LA	42227	u	**VE**	P	*VE*	NL
42175	*h	**FD**	FG	*GW*	LA	42228	m	**VE**	P	*VE*	EC
42176	*h	**FD**	FG	*GW*	LA	42229	u	**VE**	P	*VE*	NL
42177	*h	**FD**	FG	*GW*	LA	42230	u	**ST**	P	*EM*	NL
42178	*h	**FD**	P	*GW*	LA	42231	*h	**FD**	FG	*GW*	LA
42179	m	**VE**	A	*VE*	EC	42232	*h	**FD**	FG	*GW*	LA
42180	m	**VE**	A	*VE*	EC	42233	*h	**FD**	FG	*GW*	LA
42181	m	**VE**	A	*VE*	EC	42234	c	**XC**	P	*XC*	EC
42182	m	**VE**	A	*VE*	EC	42235	m	**VE**	A	*VE*	EC
42183	*d	**FD**	A	*GW*	LA	42236	*h	**FD**	A	*GW*	LA
42184	*	**FD**	A	*GW*	LA	42237	m	**VE**	P	*VE*	EC
42185	*	**FD**	A	*GW*	LA	42238	mt	**NX**	A	*VE*	EC
42186	m	**VE**	A	*VE*	EC	42239	mt	**VE**	A	*VE*	EC
42188	mt	**VE**	A	*VE*	EC	42240	m	**VE**	A	*VE*	EC
42189	mt	**VE**	A	*VE*	EC	42241	m	**VE**	A	*VE*	EC
42190	m	**VE**	A	*VE*	EC	42242	m	**VE**	A	*VE*	EC
42191	m	**NX**	A	*VE*	EC	42243	m	**VE**	A	*VE*	EC
42192	m	**NX**	A	*VE*	EC	42244	m	**VE**	A	*VE*	EC
42193	m	**NX**	A	*VE*	EC	42245	*	**FD**	A	*GW*	LA
42194	w	**VE**	P	*VE*	NL	42247	*h	**FD**	P	*GW*	LA
42195	*k	**FD**	P	*GW*	LA	42250	*	**FD**	A	*GW*	LA
42196	*h	**FD**	A	*GW*	LA	42251	*k	**FD**	A	*GW*	LA
42197	*h	**FD**	A	*GW*	LA	42252	*	**FD**	A	*GW*	LA
42198	m	**FD**	A	*GW*	LA	42253	*	**FD**	A	*GW*	LA
42199	m	**VE**	A	*VE*	EC	42255	*d	**FD**	A	*GW*	LA
42200	*d	**FD**	A	*GW*	LA	42256	*	**FD**	A	*GW*	LA
42201	*d	**FD**	A	*GW*	LA	42257	*	**FD**	A	*GW*	LA
42202	*k	**FD**	A	*GW*	LA	42258	*h	**FD**	P	*GW*	LA
42203	*h	**FD**	A	*GW*	LA	42259	*k	**FD**	A	*GW*	LA
42204	*h	**FD**	A	*GW*	LA	42260	*h	**FD**	A	*GW*	LA
42205		**EC**	P	*VE*	EC	42261	*h	**FD**	A	*GW*	LA
42206	*d	**FD**	A	*GW*	LA	42263	*	**FD**	A	*GW*	LA
42207	*d	**FD**	A	*GW*	LA	42264	*k	**FD**	A	*GW*	LA
42208	*	**FD**	A	*GW*	LA	42265	*	**FD**	A	*GW*	LA
42209	*	**FD**	A	*GW*	LA	42266	*k	**FD**	P	*GW*	LA
42210		**EC**	P	*VE*	EC	42267	*d	**FD**	A	*GW*	LA
42211	*k	**FD**	A	*GW*	LA	42268	*d	**FD**	A	*GW*	LA

42269	*	**FD**	A	*GW*	LA		42302	*k	**FD**	FG	*GW*	LA
42271	*k	**FD**	A	*GW*	LA		42303	*h	**FD**	FG	*GW*	LA
42272	*h	**FD**	A	*GW*	LA		42304	*h	**FD**	FG	*GW*	LA
42273	*h	**FD**	A	*GW*	LA		42305	*h	**FD**	FG	*GW*	LA
42275	*d	**FD**	A	*GW*	LA		42306	m	**VE**	P	*VE*	EC
42276	*	**FD**	A	*GW*	LA		42307	m	**VE**	P	*VE*	EC
42277	*	**FD**	A	*GW*	LA		42308	*h	**FD**	P	*GW*	LA
42279	*d	**FD**	A	*GW*	LA		42310	*k	**FD**	P	*GW*	LA
42280	*	**FD**	A	*GW*	LA		42315	*h	**FD**	P	*GW*	LA
42281	*	**FD**	A	*GW*	LA		42317	*k	**FD**	P	*GW*	LA
42283	*h	**FD**	A	*GW*	LA		42319	*h	**FD**	P	*GW*	LA
42284	*h	**FD**	A	*GW*	LA		42321	*h	**FD**	P	*GW*	LA
42285	*h	**FD**	A	*GW*	LA		42322	m	**VE**	P	*VE*	EC
42286	m†	**VE**	P	*VE*	EC		42323	m	**NX**	A	*VE*	EC
42287	*k	**FD**	A	*GW*	LA		42325	*	**FD**	A	*GW*	LA
42288	*h	**FD**	A	*GW*	LA		42326	m	**VE**	P	*VE*	EC
42289	*h	**FD**	A	*GW*	LA		42327	w	**ST**	P	*EM*	NL
42290	c	**XC**	P	*XC*	EC		42328	w	**ST**	P	*EM*	NL
42291	*d	**FD**	A	*GW*	LA		42329	w	**ST**	P	*EM*	NL
42292	*d	**FD**	A	*GW*	LA		42330	m	**VE**	P	*VE*	EC
42293	*	**FD**	A	*GW*	LA		42331	u	**ST**	P	*EM*	NL
42294	h	**FD**	P	*GW*	LA		42332	*	**FD**	A	*GW*	LA
42295	*d	**FD**	A	*GW*	LA		42333	*	**FD**	A	*GW*	LA
42296	*	**FD**	A	*GW*	LA		42335	w	**EC**	P	*VE*	EC
42297	*	**FD**	A	*GW*	LA		42337	w	**ST**	P	*EM*	NL
42299	*d	**GW**	A	*GW*	LA		42339	w	**ST**	P	*EM*	NL
42300	*	**GW**	A	*GW*	LA		42340	m	**VE**	A	*VE*	EC
42301	*	**GW**	A	*GW*	LA		42341	u	**ST**	P	*EM*	NL

42342	(44082)		c	**XC**	A	*XC*	EC
42343	(44095)		*	**FD**	A	*GW*	LA
42344	(44092)		*k	**FD**	A	*GW*	LA
42345	(44096)		*d	**FD**	A	*GW*	LA
42346	(41053)		*h	**FD**	A	*GW*	LA
42347	(41054)		*k	**FD**	A	*GW*	LA
42348	(41073)		*k	**FD**	A	*GW*	LA
42349	(41074)		*h	**FD**	A	*GW*	LA
42350	(41047)		*	**FD**	A	*GW*	LA
42351	(41048)		*	**GW**	A	*GW*	LA
42353	(42001, 41171)		*k	**FD**	FG	*GW*	LA
42354	(42114, 41175)		m	**VE**	A	*VE*	EC
42355	(42000, 41172)		m	**VE**	A	*VE*	EC
42356	(42002, 41173)		*k	**FD**	A	*GW*	LA
42357	(41002, 41174)		m	**VE**	A	*VE*	EC
42360	(44084, 45084)		*h	**FD**	A	*GW*	LA
42361	(44099, 42000)		*h	**FD**	A	*GW*	LA
42362	(42011, 41178)		*h	**FD**	A	*GW*	LA
42363	(41082)		m†	**VE**	A	*VE*	EC
42364	(41080)		*k	**FD**	P	*GW*	LA
42365	(41107)		*h	**FD**	P	*GW*	LA

42366–378 were converted from loco-hauled Mark 3 vehicles for CrossCountry and 42382/383 for First Great Western.

42366	(12007)	§c **XC**	P	*XC*	EC
42367	(12025)	c **XC**	P	*XC*	EC
42368	(12028)	c **XC**	P	*XC*	EC
42369	(12050)	c **XC**	P	*XC*	EC
42370	(12086)	c **XC**	P	*XC*	EC
42371	(12052)	§c **XC**	P	*XC*	EC
42372	(12055)	c **XC**	P	*XC*	EC
42373	(12071)	c **XC**	P	*XC*	EC
42374	(12075)	c **XC**	P	*XC*	EC
42375	(12113)	c **XC**	P	*XC*	EC
42376	(12085)	§c **XC**	P	*XC*	EC
42377	(12102)	c **XC**	P	*XC*	EC
42378	(12123)	c **XC**	P	*XC*	EC
42379	(41036)	§c **XC**	A	*XC*	EC
42380	(41025)	§c **XC**	A	*XC*	EC
42381	(41058)	*k **FD**	P	*GW*	LA
42382	(12128)	*h **FD**	P	*GW*	LA
42383	(12172)	*h **FD**	P	*GW*	LA
42384	(41078)	w **ST**	P	*EM*	NL
42385	(41153)	*h **FD**	P	*GW*	LA

The following carriages have been converted from loco-hauled Mark 3 vehicles for Grand Central. They have a lower density seating layout (most seats arranged around tables). –/64 2T († –/60 1TD 1T 2W).

42401	(12149)	**GC**	A	*GC*	HT
42402	(12155)	**GC**	A	*GC*	HT
42403	(12033) †	**GC**	A	*GC*	HT
42404	(12152)	**GC**	A	*GC*	HT
42405	(12136)	**GC**	A	*GC*	HT
42406	(12112) †	**GC**	A	*GC*	HT
42407	(12044)	**GC**	A	*GC*	HT
42408	(12121)	**GC**	A	*GC*	HT
42409	(12088) †	**GC**	A	*GC*	HT

These carriages have been converted from TFKB or TSB buffet cars to TS vehicles in 2011–12 (42501–515) and 2013–14 (42516–520) at Wabtec Kilmarnock for FGW. Refurbished with Grammer seating.–/84 1T. 34.8 t.

42501	(40344, 40744)	**FD**	A	*GW*	LA
42502	(40331, 40731)	**FD**	A	*GW*	LA
42503	(40312, 40712)	**FD**	A	*GW*	LA
42504	(40314, 40714)	**FD**	A	*GW*	LA
42505	(40428, 40228)	**FD**	A	*GW*	LA
42506	(40324, 40724)	**FD**	A	*GW*	LA
42507	(40409, 40209)	**FD**	A	*GW*	LA
42508	(40325, 40725)	**FD**	A	*GW*	LA
42509	(40336, 40736)	**FD**	A	*GW*	LA
42510	(40317, 40717)	**FD**	A	*GW*	LA
42511	(40309, 40709)	**FD**	A	*GW*	LA

42512	(40408, 40208)	**FD**	A	*GW*	LA
42513	(40338, 40738)	**FD**	A	*GW*	LA
42514	(40326, 40726)	**FD**	A	*GW*	LA
42515	(40347, 40747)	**FD**	A	*GW*	LA
42516	(40323, 40723)	**FD**	A	*GW*	LA
42517	(40345, 40745)	**FD**	A	*GW*	LA
42518	(40003, 40403)	**FD**	P	*GW*	LA
42519	(40016, 40416)	**FD**	P	*GW*	LA
42520	(40234, 40434)	**FD**	P	*GW*	LA

These carriages have been converted from TF to TS vehicles in 2014 at Wabtec Kilmarnock for FGW. Refurbished with Grammer seating. –/80 1T. 35.5 t.

42551	(41003)	**FD**	A	*GW*	LA
42552	(41007)	**FD**	A	*GW*	LA
42553	(41009)	**FD**	A	*GW*	LA
42554	(41011)	**FD**	A	*GW*	LA
42555	(41015)	**FD**	A	*GW*	LA
42556	(41017)	**FD**	A	*GW*	LA
42557	(41019)	**FD**	A	*GW*	LA
42558	(41021)	**FD**	A	*GW*	LA
42559	(41023)	**FD**	A	*GW*	LA
42560	(41027)	**FD**	A	*GW*	LA
42561	(41031)	**FD**	A	*GW*	LA
42562	(41037)	**FD**	A	*GW*	LA
42563	(41045)	**FD**	FG	*GW*	LA
42564	(41051)	**FD**	A	*GW*	LA
42565	(41085)	**FB**	FG	*GW*	LA
42566	(41086)	**FD**	FG	*GW*	LA
42567	(41093)	**FD**	A	*GW*	LA
42568	(41101)	**FD**	A	*GW*	LA
42569	(41105)	**FD**	A	*GW*	LA
42570	(41114)	**FD**	FG	*GW*	LA
42571	(41121)	**FD**	A	*GW*	LA
42572	(41123)	**FD**	A	*GW*	LA
42573	(41127)	**FD**	A	*GW*	LA
42574	(41129)	**FD**	A	*GW*	LA
42575	(41131)	**FD**	A	*GW*	LA
42576	(41133)	**FD**	A	*GW*	LA
42577	(41141)	**FD**	A	*GW*	LA
42578	(41143)	**FD**	A	*GW*	LA
42579	(41145)	**GW**	A	*GW*	LA
42580	(41155)	**FD**	P	*GW*	LA
42581	(41157)	**FD**	A	*GW*	LA
42582	(41163)	**FD**	FG	*GW*	LA
42583	(41153, 42385)	**FD**	P		

TRAILER GUARD'S STANDARD   TGS

As built and m –/65 1T.
* Refurbished Great Western Railway vehicles. Grammer seating and toilet removed for trolley store. –/67 (unless h).

44000–44083

- t Contains a mix of original seats and 20 new Grammer seats. –/65 1T.
- c Refurbished CrossCountry vehicles with Primarius seating. –/67 1T.
- h "High density" Great Western Railway vehicles. –/71.
- m Refurbished Virgin Trains East Coast vehicles with Primarius seating.
- s Fitted with centre luggage stack (EMT) –/63 1T.
- t Fitted with centre luggage stack –/61 1T.

44000. Lot No. 30953 Derby 1980. 33.47 t.
44001–090. Lot No. 30949 Derby 1980–82. 33.47 t.
44091–094. Lot No. 30964 Derby 1982. 33.47 t.
44097–101. Lot No. 30970 Derby 1982. 33.47 t.

44000	*h	**FD**	P	*GW*	LA	44041	s	**ST**	P	*EM*	NL
44001	*	**FD**	A	*GW*	LA	44042	*h	**FD**	P	*GW*	LA
44002	*h	**FD**	A	*GW*	LA	44043	*h	**FD**	A	*GW*	LA
44003	*h	**FD**	A	*GW*	LA	44044	s	**ST**	P	*EM*	NL
44004	*h	**FD**	A	*GW*	LA	44045	m	**VE**	A	*VE*	EC
44005	*h	**FD**	A	*GW*	LA	44046	s	**ST**	P	*EM*	NL
44007	*h	**FD**	A	*GW*	LA	44047	s	**ST**	P	*EM*	NL
44008	*h	**FD**	A	*GW*	LA	44048	s	**ST**	P	*EM*	NL
44009	*h	**FD**	A	*GW*	LA	44049	*	**FD**	A	*GW*	LA
44010	*h	**FD**	A	*GW*	LA	44050	m	**VE**	P	*VE*	EC
44011	*	**FD**	A	*GW*	LA	44051	s	**ST**	P	*EM*	NL
44012	c	**XC**	A	*XC*	EC	44052	c	**XC**	P	*XC*	EC
44013	*h	**FD**	A	*GW*	LA	44054	s	**ST**	P	*EM*	NL
44014	*h	**FD**	A	*GW*	LA	44055	*h	**FB**	FG	*GW*	LA
44015	*	**FD**	A	*GW*	LA	44056	m	**VE**	A	*VE*	EC
44016	*h	**FD**	A	*GW*	LA	44057	m	**NX**	P	*VE*	EC
44017	c	**XC**	A	*XC*	EC	44058	m	**VE**	A	*VE*	EC
44018	*	**FD**	A	*GW*	LA	44059	*	**FD**	A	*GW*	LA
44019	m	**VE**	A	*VE*	EC	44060	*h	**FD**	P	*GW*	LA
44020	*h	**FD**	A	*GW*	LA	44061	m	**NX**	A	*VE*	EC
44021	c	**XC**	P	*XC*	EC	44063	m	**VE**	A	*VE*	EC
44022	*h	**FD**	A	*GW*	LA	44064	*h	**FD**	A	*GW*	LA
44023	*h	**FD**	A	*GW*	LA	44065	t	**V**	AV		LM
44024	*h	**FD**	A	*GW*	OO	44066	*	**FD**	A	*GW*	LA
44025	*	**FD**	A	*GW*	LA	44067	*h	**FD**	A	*GW*	LA
44026	*h	**FD**	A	*GW*	LA	44068	*h	**FD**	FG	*GW*	LA
44027	s	**VE**	P	*VE*	NL	44069	*h	**FD**	P	*GW*	LA
44028	*	**FD**	A	*GW*	LA	44070	s	**ST**	P	*EM*	NL
44029	*	**FD**	A	*GW*	LA	44071	s	**ST**	P	*EM*	NL
44030	*h	**FD**	A	*GW*	LA	44072	c	**XC**	P	*XC*	EC
44031	m	**VE**	A	*VE*	EC	44073	t	**EC**	P	*VE*	EC
44032	*	**FD**	A	*GW*	LA	44074	*h	**FD**	FG	*GW*	LA
44033	*h	**FD**	A	*GW*	LA	44075	m	**VE**	P	*VE*	EC
44034	*	**FD**	A	*GW*	LA	44076	*h	**FD**	FG	*GW*	LA
44035	*	**FD**	A	*GW*	LA	44077	m	**VE**	A	*VE*	EC
44036	*h	**FD**	A	*GW*	LA	44078	*h	**FD**	P	*GW*	LA
44037	*h	**FD**	A	*GW*	LA	44079	*h	**FD**	P	*GW*	LA
44038	*	**FD**	A	*GW*	LA	44080	m	**VE**	A	*VE*	EC
44039		**FD**	A	*GW*	LA	44081	*h	**FD**	FG	*GW*	LA
44040	*	**GW**	A	*GW*	LA	44083	*h	**FD**	P	*GW*	LA

44085	s	**ST**	P	*EM*	NL	44093	*h	**FD**	A	*GW*	LA
44086	*	**FD**	A	*GW*	LA	44094	m	**VE**	A	*VE*	EC
44088	t	**V**	AV		LM	44097	*h	**FD**	P	*GW*	LA
44089	t	**V**	AV		LM	44098	m	**VE**	A	*VE*	EC
44090	*h	**FD**	P	*GW*	LA	44100	*h	**FD**	FG	*GW*	PM
44091	*h	**FD**	P	*GW*	LA	44101	*h	**FD**	P	*GW*	LA

TRAILER COMPOSITE KITCHEN   TCK

Converted from Mark 3A Open Standard. Refurbished CrossCountry vehicles with Primarius seating. Small kitchen for the preparation of hot food and stowage space for two trolleys between First and Standard Class. One toilet removed. 30/10 1T.

45001–005. Lot No. 30877 Derby 1975–77. 34.3 t.

45001	(12004)	**XC**	P	*XC*	EC
45002	(12106)	**XC**	P	*XC*	EC
45003	(12076)	**XC**	P	*XC*	EC
45004	(12077)	**XC**	P	*XC*	EC
45005	(12080)	**XC**	P	*XC*	EC

TRAILER COMPOSITE   TC

Converted from TF for First Great Western 2014–15. Refurbished with Grammer seating. 24/39 1T.

46001–004. Lot No. 30881 Derby 1976–77.   t. Converted from TF.
46005–009. Lot No. 30896 Derby 1977–78.   t. Converted from TF.
46010–013. Lot No. 30938 Derby 1979–80.   t. Converted from TF.
46014. Lot No. 30963 Derby 1982.   t. Converted from TF.
46015. Lot No. 30884 Derby 1976–77.   t. Converted from TF.
46016/017. Lot No. 30939 Derby 1979–80.   t. Converted from TF.
46018. Lot No. 30969 Derby 1982.   t. Converted from TF.

46001	(41005)	**FD**	A	*GW*	LA
46002	(41029)	**FD**	A	*GW*	LA
46003	(41033)	**FD**	A	*GW*	LA
46004	(41055)	**FD**	A	*GW*	LA
46005	(41065)	**FD**	A	*GW*	LA
46006	(41081)	**FD**	P	*GW*	LA
46007	(41096)	**FD**	P	*GW*	LA
46008	(41109)	**FD**	P	*GW*	LA
46009	(41119)	**FD**	P	*GW*	LA
46010	(41125)	**FD**	A	*GW*	LA
46011	(41139)	**FD**	A	*GW*	LA
46012	(41147)	**FD**	P	*GW*	LA
46013	(41148)	**FD**	P	*GW*	LA
46014	(41168)	**FD**	P	*GW*	LA
46015	(40505, 41179)	**FD**	A	*GW*	LA
46016	(42282, 41181)	**FD**	P	*GW*	LA
46017	(42270, 41184)	**FD**	P	*GW*	LA
46018	(42318, 41191)	**FD**	P	*GW*	LA

PLATFORM 5 MAIL ORDER
EUROPEAN HANDBOOKS

The Platform 5 European Railway Handbooks are the most comprehensive guides to the rolling stock of selected European railway administrations available. Each book lists all locomotives and railcars of the country concerned, giving details of number carried and depot allocation, together with a wealth of technical data for each class of vehicle. Each book is A5 size, thread sewn and illustrated throughout with colour photographs. The Irish book also contains details of hauled coaching stock.

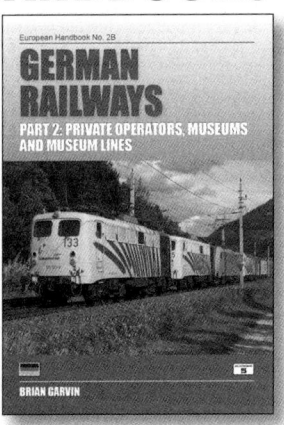

EUROPEAN HANDBOOKS CURRENTLY AVAILABLE:

No. 1 Benelux Railways (2012)	£20.95
No. 2A German Railways Part 1: DB Locomotives & Multiple Units (2013)	£22.95
No. 2B German Railways Part 2: Private Operators, Museums & Museum Lines (2015)	£26.95
No. 3 Austrian Railways (2012)	£19.95
No. 4 French Railways (2011)	£19.95
No. 5 Swiss Railways (2009)	£19.50
No. 6 Italian Railways (2014)	£21.95
No. 7 Irish Railways (2013)	£15.95

Please add postage: 10% UK, 20% Europe, 30% Rest of World.

Telephone, fax or send your order to the Platform 5 Mail Order Department. See inside back cover of this book for details.

HST SET FORMATIONS

GREAT WESTERN RAILWAY

The largest operator of HSTs is Great Western Railway with 53 sets to cover 49 diagrams (of these one is a "hot spare" at Old Oak Common and one a "hot spare" at Bristol St Philip's Marsh).

The sets are split into three types, as shown below. These are 16 "low density" sets mainly used on West Country services, 19 "high density" sets with a full kitchen or buffet vehicle and 18 "super high density" sets including a TSB vehicle (with just a small corner buffet counter).

Although some sets are prefixed "OC" for Old Oak Common, for maintenance purposes all trailers are now based at Laira apart from two spare vehicles.

By summer 2015 all GWR HSTs were 8-car rakes with one full First Class carriage and either a First Class buffet car or one of the newly converted Composite vehicles – essentially both providing 1½ coaches of First Class.

Number of sets: 53.
Maximum number of daily diagrams: 49.
Formations: 8-cars.
Allocation: Laira (Plymouth).
Other maintenance and servicing depots: Landore (Swansea), St Philip's Marsh (Bristol), Long Rock (Penzance).
Operation: London Paddington–Exeter/Paignton/Plymouth/(Newquay in the summer)/Penzance, Bristol, Cardiff/Swansea/West Wales, Oxford/Hereford/Great Malvern/Cheltenham Spa.

Set	L	K	F	E	D	C	B	A	density
LA01	41024	40755	42034	42559	42033	42007	42035	44011	L
LA02	41032	40727	42046	42561	42045	42202	42047	44015	L
LA03	41038	40757	42055	42562	42343	42292	42056	44018	L
LA04	41052	40710	42077	42564	42076	42004	42078	44025	L
LA05	41094	40707	42185	42567	42184	42183	42107	44001	L
LA06	41104	40713	42208	42505	42054	42206	42209	44066	L
LA07	41122	40722	42252	42571	42019	42345	42253	44028	L
LA08	41124	40739	42256	42572	42263	42255	42257	44029	L
LA09	41130	40716	42072	42574	42325	42267	42269	44032	L
LA10	41134	40721	42276	42576	42332	42275	42277	44034	L
LA11	41135	40703	42280	42580	42265	42279	42281	44035	L
LA12	41136	40752	42144	42551	42143	42268	42145	44049	L
LA13	41142	40734	42333	42577	42075	42291	42293	44038	L
LA14	41144	40733	42296	42578	42350	42295	42297	44039	L
LA15	41146	40715	42300	42579	42351	42299	42301	44040	L
LA16	41158	40743	42245	42581	42129	42200	42250	44086	L
OC30	41008	40807	42079	42552	42236	42251	42080	44026	H
OC31	41128	40804	42060	42573	42197	42347	42061	44020	H
OC32	41018	40801	42025	42556	42362	42024	42026	44008	H
OC33	41028	40806	42040	42560	42039	42348	42041	44013	H

HST SET FORMATIONS

Set									density
OC34	41102	40803	42203	42568	42027	42201	42204	44064	H
OC35	41106	40808	42213	42569	42212	42211	42214	44067	H
OC36	41110	40809	42346	42511	42349	42138	42089	44003	H
OC37	41132	40810	42272	42575	42073	42271	42273	44033	H
OC38	41138	40810	42284	42517	42003	42202	42285	44036	H
LA60	41162	40900	42231	42570	42232	42353	42233	44074	H
LA61	41166	40901	42304	42566	42303	42302	42305	44068	H
LA62	41167	40902	42167	42565	42168	42103	42169	44055	H
LA63	41059	40903	42175	42563	42176	42105	42177	44081	H
LA64	41166	40904	42094	42582	42093	42092	42108	44076	H
LA71	41010	40204	42013	42553	42360	42012	42014	44004	H
LA72	41012	40205	42016	42554	42005	42015	42361	44005	H
LA73	41016	40207	42006	42555	42096	42021	42023	44007	H
LA74	41020	40221	42028	42557	42009	42259	42029	44009	H
LA75	41022	40210	42031	42558	42010	42030	42032	44010	H

Set	L	K	F	E	D	C	B	A	density
OC40	41149	46016	40106	42502	42166	42218	42071	44079	SH
OC41	41182	46008	40107	42520	42222	42224	42382	44101	SH
OC42	41183	46014	40108	42519	42315	42317	42383	44090	SH
OC43	41186	46012	40109	42515	42258	42266	42365	44000	SH
OC44	41192	46018	40110	42514	42115	42174	42288	44097	SH
OC45	41187	46009	40111	42501	42247	42173	42260	44078	SH
OC46	41161	46007	40112	42512	42178	42195	42043	44042	SH
OC47	41108	46006	40113	42509	42308	42217	42067	44060	SH
OC48	41189	46017	40114	42518	42085	42310	42087	44069	SH
OC49	41169	46013	40115	42385	42319	42364	42321	44091	SH
OC50	41180	46015	40101	42516	42098	42264	42283	44093	SH
OC51	41006	46001	40103	42513	42070	42069	42118	44023	SH
OC52	41030	46002	40102	42510	42042	42008	42044	44014	SH
OC53	41034	46003	40118	42506	42048	42066	42050	44016	SH
OC54	41056	46004	40117	42503	42074	42081	42126	44043	SH
OC55	41089	46005	40104	42504	42221	42062	42068	44022	SH
OC56	41126	46010	40116	42508	42216	42344	42261	44030	SH
OC57	41140	46011	40105	42507	42099	42287	42289	44037	SH

L = Low density; H = High density; SH = Super High density.

Spares:

LA: 40119 40231 40718 40811 41004 41103 41116 41137 41176
42049 42083 42095 42101 42102 42196 42294 42356 42381
44002 44059 44083
OO: 44024
PM: 44100

HST SET FORMATIONS

VIRGIN TRAINS EAST COAST

Virgin Trains East Coast operates 15 refurbished HST sets on the ECML. As well as serving non-electrified destinations such as Hull and Inverness the VTEC HSTs also work alongside Class 91s and Mark 4 sets on services to Leeds, Newcastle and Edinburgh. Set EC64 (ex-NL05) transferred from East Midlands Trains in 2011 and set EC65 (ex-NL02) as an 8-car followed in 2015.

Number of sets: 15.
Maximum number of daily diagrams: 14.
Formations: 9-cars.
Allocation: Craigentinny (Edinburgh).
Other maintenance depot: Neville Hill (Leeds).
Operation: London King's Cross–Leeds/Harrogate/Skipton/Hull/Lincoln/Newcastle/Edinburgh/Aberdeen/Inverness.

Set	M	L	J	G	F	E	D	C	B
EC51	41120	41150	40748	42215	42091	42146	42150	42154	44094
EC52	41039	41040	40735	42323	42189	42057	42058	42059	44019
EC53	41090	41044	40737	42340	42127	42063	42064	42065	44045
EC54	41087	41088	40706	42104	42161	42171	42172	42219	44056
EC55	41091	41092	40704	42179	42188	42180	42181	42106	44058
EC56	41170	41118	40720	42241	42363	42242	42243	42244	44098
EC57	41151	41152	40740	42226	42128	42182	42186	42190	44080
EC58	41097	41098	40750	42158	42238	42191	42192	42193	44061
EC59	41099	41100	40711	42235	42239	42240	42198	42199	44063
EC60	41066	41164	40742	42122	42116	42357	42134	42355	44031
EC61	41115	41165	40702	42117	42159	42160	42109	42110	44057
EC62	41185	41095	40701	42306	42326	42330	42237	42307	44075
EC63	41159	41083	40708	42163	42286	42228	42130	42322	44069
EC64	41062	41154	40732	42125	42335	42123	42205	42210	44073
EC65	41112	41068	40751	42194	42229	42227	42225		44027

Spares:
EC: 40705 41190 42354 44077

GRAND CENTRAL

GC operates HSTs between Sunderland and King's Cross. Formations are flexible: there are enough vehicles to form three 6-car sets or one set can be disbanded to enable the sets to run as 7-cars at times of high demand.

Number of sets: 3.
Maximum number of daily diagrams: 2.
Formations: 6-cars or 7-cars.
Allocation: Heaton (Newcastle).
Operation: London King's Cross–Sunderland.

Set	TF	TSB	TS	TS	TS	TF*
GC01	41201	40424	42403	42402	42401	41204
GC02	41202	40426	42406	42405	42404	41205
GC03	41203	40433	42409	42408	42407	41206

* declassified.

HST SET FORMATIONS

EAST MIDLANDS TRAINS

East Midlands Trains HSTs are concentrated on the Nottingham corridor during the day, with early morning and evening services to Leeds for servicing at Neville Hill.

Number of sets: 9.
Maximum number of daily diagrams: 8.
Formations: 8-cars.
Allocation: Neville Hill (Leeds).
Other maintenance depot: Derby Etches Park.
Operation: London St Pancras–Nottingham, Sheffield/Leeds.

Set	J	G	F	E	D	C	B	A
NL01	41057	41084	40730	42327	42111	42112	42113	44041
NL03	41061	41067	40741	42337	42119	42120	42121	44054
NL04	41077	41064	40749	42151	42164	42165	42153	44047
NL06	41156	41041	40746	42132	42131	42331	42133	44046
NL07	41111	41070	40754	42339	42135	42136	42137	44044
NL08	41071	41072	40753	42329	42139	42140	42141	44048
NL10	41075	41076	40756	42328	42341	42148	42149	44051
NL11	41117	41046	40728	42220	42100	42230	42124	44085
NL12	41079	41069	40700	42155	42156	42157	42152	44070

Spares:

NL: 41063 41113 42384 44071

CROSSCOUNTRY

CrossCountry reintroduced regular HST diagrams on its services from the December 2008 timetable. Trains run in 7-car formation with the spare TS coaches regularly used in traffic as coaches "C", "D" or "E".

Number of sets: 5.
Maximum number of daily diagrams: 4.
Formations: 7-cars.
Allocation: Craigentinny (Edinburgh).
Other maintenance depots: Laira (Plymouth) or Neville Hill (Leeds).
Operation: Edinburgh–Leeds–Plymouth is the core route with some services extending to Dundee or Penzance.

Set	A	B	C	D	E	F	G
XC01	41193	45001	42368	42369	42367	42366	44021
XC02	41194	45002	42375	42373	42371	42371	44072
XC03	41195	45003	42290	42378	42377	42376	44052
XC04	41026	45004	42038	42037	42036	42380	44012
XC05	41035	45005	42051	42053	42052	42379	44017

Spares:

EC: 42097 42234 42342 42370 42372

3. SALOONS

Several specialist passenger carrying carriages, normally referred to as saloons are permitted to run on the national railway system. Many of these are to pre-nationalisation designs.

WCJS FIRST CLASS SALOON

Built 1892 by LNWR, Wolverton. Originally dining saloon mounted on six-wheel bogies. Rebuilt with new underframe with four-wheel bogies in 1927. Rebuilt 1960 as observation saloon with DMU end. Gangwayed at other end. The interior has a saloon, kitchen, guards vestibule and observation lounge. 19/– 1T. Gresley bogies. 28.5 t. 75 mph. ETS x.

41 (484, 45018) x **M** WC *WC* CS

LNWR DINING SALOON

Built 1890 by LNWR, Wolverton. Mounted on the underframe of LMS General Utility Van 37908 in the 1980s. Contains kitchen and dining area seating 12 at tables for two. 12/–. Gresley bogies. 75 mph. 25.4 t. ETS x.

159 (5159) x **M** WC *WC* CS

GNR FIRST CLASS SALOON

Built 1912 by GNR, Doncaster. Contains entrance vestibule, lavatory, two separate saloons, library and luggage space. 19/– 1T. Gresley bogies. 75 mph. 29.4 t. ETS x.

Non-standard livery: Teak.

807 (4807) x **0** WC *WC* CS

LNER GENERAL MANAGERS SALOON

Built 1945 by LNER, York. Gangwayed at one end with a veranda at the other. The interior has a dining saloon seating 12, kitchen, toilet, office and nine seat lounge. 21/– 1T. B4 bogies. 75 mph. 35.7 t. ETS 3.

1999 (902260) **M** BE CS DINING CAR No. 2

GENERAL MANAGER'S SALOON

Renumbered 1989 from London Midland Region departmental series. Formerly the LMR General Manager's saloon. Rebuilt from LMS period 1 Corridor Brake First M5033M to dia 1654 and mounted on the underframe of BR suburban Brake Standard M43232. Screw couplings have been removed. B4 bogies. 100 mph. ETS 2X.

LMS Lot No. 326 Derby 1927. 27.5 t.

6320 (5033, DM 395707)   x   **M** PR *PR* SK

BELMOND BRITISH PULLMAN SUPPORT CAR

Converted 199? from Courier vehicle converted from Mark 1 Corridor Brake Standard 1986–87. Toilet retained and former compartment area replaced with train manager's office, crew locker room, linen store and dry goods store. The former luggage area has been adapted for use as an engineers' compartment and workshop. Commonwealth bogies. 100 mph. ETS 2.

Lot No. 30721 Wolverton 1963. 37 t.

99545 (35466, 80207)   **PC** BE *BP* SL   BAGGAGE CAR No. 11

SERVICE CAR

Converted from BR Mark 1 Corridor Brake Standard. Commonwealth bogies. 100 mph. ETS 2.

Lot No. 30721 Wolverton 1963.

99886 (35407)   x   **M** WC *WC* CS   SERVICE CAR No. 86

ROYAL SCOTSMAN SALOONS

Built 1960 by Metro-Cammell as Pullman Kitchen First for East Coast Main Line services. Rebuilt 2013 as dining car. Commonwealth bogies. 38.5 t. ETS x.

99960 (321 SWIFT)   **M** BE *RS* CS   DINING CAR No. 2

Built 1960 by Metro-Cammell as Pullman Parlour First (§ Pullman Kitchen First) for East Coast Main Line services. Rebuilt 1990 as sleeping cars with four twin sleeping rooms (*§ three twin sleeping rooms and two single sleeping rooms at each end). Commonwealth bogies. 38.5 t. ETS x.

99961 (324 AMBER) *   **M** BE *RS* CS   STATE CAR 1
99962 (329 PEARL)   **M** BE *RS* CS   STATE CAR 2
99963 (331 TOPAZ)   **M** BE *RS* CS   STATE CAR 3
99964 (313 FINCH) §   **M** BE *RS* CS   STATE CAR 4

Built 1960 by Metro-Cammell as Pullman Kitchen First for East Coast Main Line services. Rebuilt 1990 as observation car with open verandah seating 32. Commonwealth bogies. 38.5 t. ETS x.

| 99965 (319 SNIPE) | M | BE | *RS* | CS | OBSERVATION CAR |

Built 1960 by Metro-Cammell as Pullman Kitchen First for East Coast Main Line services. Rebuilt 1993 as dining car. Commonwealth bogies. 38.5 t. ETS x.

| 99967 (317 RAVEN) | M | BE | *RS* | CS | DINING CAR No. 1 |

Mark 3A. Converted 1997 from a Sleeping Car at Carnforth Railway Restoration & Engineering Services. BT10 bogies. Attendant's and adjacent two sleeping compartments converted to generator room containing a 160 kW Volvo unit. In 99968 four sleeping compartments remain for staff use with another converted for use as a staff shower and toilet. The remaining five sleeping compartments have been replaced by two passenger cabins. In 99969 seven sleeping compartments remain for staff use. A further sleeping compartment, along with one toilet, have been converted to store rooms. The other two sleeping compartments have been combined to form a crew mess. 41.5 t. ETS 7X.

Lot No. 30960 Derby 1981–83.

| 99968 (10541) | M | BE | *RS* | CS | STATE CAR 5 |
| 99969 (10556) | M | BE | *RS* | CS | SERVICE CAR |

RAILFILMS "LMS CLUB CAR"

Converted from BR Mark 1 Open Standard at Carnforth Railway Restoration & Engineering Services in 1994. Contains kitchen, pantry and two dining saloons. 20/– 1T. Commonwealth bogies. 100 mph. ETS 4.

Lot No. 30724 York 1963. 37 t.

| 99993 (5067) | x | M | RF | *ST* | CS | LMS CLUB CAR |

BR INSPECTION SALOON

Mark 1. Short frames. Non-gangwayed. Observation windows at each end. The interior layout consists of two saloons interspersed by a central lavatory/kitchen/guards/luggage section. 90 mph. ETS x.

BR Wagon Lot No. 3095 Swindon 1957. B4 bogies. 30.5 t.

| 999506 | M | WC | *WC* | CS |

4. PULLMAN CAR COMPANY SERIES

Pullman cars have never generally been numbered as such, although many have carried numbers, instead they have carried titles. However, a scheme of schedule numbers exists which generally lists cars in chronological order. In this section those numbers are shown followed by the car's title. Cars described as "kitchen" contain a kitchen in addition to passenger accommodation and have gas cooking unless otherwise stated. Cars described as "parlour" consist entirely of passenger accommodation. Cars described as "brake" contain a compartment for the use of the guard and a luggage compartment in addition to passenger accommodation.

PULLMAN PARLOUR FIRST

Built 1927 by Midland Carriage & Wagon Company. 26/– 2T. Gresley bogies. 41 t. ETS 2.

| 213 | MINERVA | **PC** | BE | *BP* | SL |

PULLMAN KITCHEN FIRST

Built 1928 by Metropolitan Carriage & Wagon Company. 20/– 1T. Gresley bogies. 42 t. ETS 4.

| 238 | PHYLISS | **PC** | BE | | SL |

PULLMAN PARLOUR FIRST

Built 1928 by Metropolitan Carriage & Wagon Company. 24/– 2T. Gresley bogies. 40 t. ETS 4.

| 239 | AGATHA | **PC** | BE | | SL |
| 243 | LUCILLE | **PC** | BE | *BP* | SL |

PULLMAN KITCHEN FIRST

Built 1925 by BRCW. Rebuilt by Midland Carriage & Wagon Company in 1928. 20/– 1T. Gresley bogies. 41 t. ETS 4.

| 245 | IBIS | **PC** | BE | *BP* | SL |

PULLMAN PARLOUR FIRST

Built 1928 by Metropolitan Carriage & Wagon Company. 24/– 2T. Gresley bogies. ETS 4.

| 254 | ZENA | **PC** | BE | *BP* | SL |

PULLMAN KITCHEN FIRST

Built 1928 by Metropolitan Carriage & Wagon Company. 20/– 1T. Gresley bogies. 42 t. ETS 4.

| 255 | IONE | | **PC** | BE | *BP* | SL |

PULLMAN KITCHEN COMPOSITE

Built 1932 by Metropolitan Carriage & Wagon Company. Originally included in 6-Pul EMU. Electric cooking. 12/16 1T. EMU bogies. ETS x.

| 264 | RUTH | | **PC** | BE | | SL |

PULLMAN KITCHEN FIRST

Built 1932 by Metropolitan Carriage & Wagon Company. Originally included in "Brighton Belle" EMUs but now used as hauled stock. Electric cooking. 20/– 1T. B5 (SR) bogies (§ EMU bogies). 44 t. ETS 2.

280	AUDREY		**PC**	BE	*BP*	SL
281	GWEN		**PC**	BE	*BP*	SL
283	MONA	§	**PC**	BE		SL
284	VERA		**PC**	BE	*BP*	SL

PULLMAN PARLOUR THIRD

Built 1932 by Metropolitan Carriage & Wagon Company. Originally included in "Brighton Belle" EMUs. –/56 2T. EMU bogies. ETS x.

Non-standard livery: BR Revised Pullman (blue & white lined out in white).

| 286 | CAR No. 86 | **O** | BE | SL |

PULLMAN BRAKE THIRD

Built 1932 by Metropolitan Carriage & Wagon Company. Originally driving motor cars in "Brighton Belle" EMUs. Traction and control equipment removed for use as hauled stock. –/48 1T. EMU bogies. ETS x.

| 292 | CAR No. 92 | **PC** | BE | SL |
| 293 | CAR No. 93 | **PC** | BE | SL |

PULLMAN PARLOUR FIRST

Built 1951 by Birmingham Railway Carriage & Wagon Company. 32/– 2T. Gresley bogies. 39 t. ETS 3.

| 301 | PERSEUS | **PC** | BE | *BP* | SL |

Built 1952 by Pullman Car Company, Preston Park using underframe and bogies from 176 RAINBOW, the body of which had been destroyed by fire. 26/– 2T. Gresley bogies. 38 t. ETS 4.

| 302 | PHOENIX | **PC** | BE | *BP* | SL |

PULLMAN 308–354

PULLMAN PARLOUR FIRST

Built 1951 by Birmingham Railway Carriage & Wagon Company. 32/– 2T. Gresley bogies. 39 t. ETS 3.

308   CYGNUS   **PC**   BE   *BP*   SL

PULLMAN BAR FIRST

Built 1951 by Birmingham Railway Carriage & Wagon Company. Rebuilt 1999 by Blake Fabrications, Edinburgh with original timber-framed body replaced by a new fabricated steel body. Contains kitchen, bar, dining saloon and coupé. Electric cooking. 14/– 1T. Gresley bogies. ETS 3.

310   PEGASUS   x   **PC**   LS   *LS*   CL

Also carries "THE TRIANON BAR" branding.

PULLMAN PALOUR FIRST

Built 1960–61 by Metro-Cammell for East Coast Main Line services. –/36 2T. Commonwealth bogies. 38.5 t. ETS x.

326   EMERALD   x   **PC**   WC   *WC*   CS

PULLMAN KITCHEN SECOND

Built 1960–61 by Metro-Cammell for East Coast Main Line services. Commonwealth bogies. –/30 1T. 40 t. ETS x.

335   CAR No. 335   x   **PC**   VT   *VT*   TM

PULLMAN PARLOUR SECOND

Built 1960–61 by Metro-Cammell for East Coast Main Line services. 347, 348 and 350 are used as Open Firsts. –/42 2T. Commonwealth bogies. 38.5 t. ETS x.

347	CAR No. 347	x	**M**	WC	*WC*	CS	
348	CAR No. 348	x	**M**	WC	*WC*	CS	
349	CAR No. 349	x	**PC**	VT	*VT*	TM	
350	CAR No. 350	x	**M**	WC	*WC*	CS	
351	CAR No. 351	x	**PC**	WC	*WC*	CS	SAPPHIRE
352	CAR No. 352	x	**PC**	WC	*WC*	CS	AMETHYST
353	CAR No. 353	x	**PC**	VT	*VT*	TM	

PULLMAN SECOND BAR

Built 1960–61 by Metro-Cammell for East Coast Main Line services. –/24+17 bar seats. Commonwealth bogies. 38.5 t. ETS x.

354   THE HADRIAN BAR   x   **PC**   WC   *WC*   CS

5. LOCOMOTIVE SUPPORT CARRIAGES

These carriages have been adapted from Mark 1s and Mark 2s for use as support carriages for heritage steam and diesel locomotives. Some seating is retained for the use of personnel supporting the locomotives operation with the remainder of the carriage adapted for storage, workshop, dormitory and catering purposes. These carriages can spend considerable periods of time off the national railway system when the locomotives they support are not being used on that system. No owner or operator details are included in this section. After the depot code, the locomotive(s) each carriage is usually used to support is given.

CORRIDOR BRAKE FIRST

Mark 1. Commonwealth bogies. ETS 2.

14007. Lot No. 30382 Swindon 1959. 35 t.
17015. Lot No. 30668 Swindon 1961. 36 t.
17025. Lot No. 30718 Swindon 1963. Metal window frames. 36 t.

14007	(14007, 17007)	x	**M**	NY	LNER 61264
17015	(14015)	x	**CC**	TM	Tyseley Locomotive Works-based locos
17025	(14025)	v	**M**	CS	LMS 45690

CORRIDOR BRAKE FIRST

Mark 2A. Pressure ventilated. B4 bogies. ETS 4.

14064. Lot No. 30775 Derby 1967–68. 32 t.
14099/17096. Lot No. 30786 Derby 1968. 32 t.

14064	(14064, 17064)	x	**M**	CS	LMS 45305/BR 70013
14099	(14099, 17099)	v	**M**	CS	LMS 45305/BR 70013
17096	(14096)		**PC**	SL	SR 35028 MERCATOR

CORRIDOR BRAKE COMPOSITE

Mark 1. ETS 2.

21096. Lot No. 30185 Metro-Cammell 1956. BR Mark 1 bogies. 32.5 t.
21232. Lot No. 30574 GRCW 1960. B4 bogies. 34 t.
21249. Lot No. 30669 Swindon 1961–62. Commonwealth bogies. 36 t.

21096	x	**M**	NY	LNER 60007
21232	x	**M**	SK	LMS 46233
21249	x	**CC**	BH	New Build 60163

LOCOMOTIVE SUPPORT CARRIAGES

CORRIDOR BRAKE STANDARD

Mark 1. Metal window frames and melamine interior panelling. ETS 2.

35317–322. Lot No. 30699 Wolverton 1962–63. Commonwealth bogies. 37 t.
35449. Lot No. 30728 Wolverton 1963. Commonwealth bogies. 37 t.
35451–486. Lot No. 30721 Wolverton 1963. Commonwealth bogies. 37 t.

35317	x	**CC**	SH	BR 70000
35322	x	**M**	CS	WCRC Carnforth-based locomotives
35449	x	**M**	BQ	LMS 45231
35451	x	**CC**	SH	LMS 46100
35461	x	**CH**	CL	GWR 5029
35463	v	**M**	CS	WCRC Carnforth-based locomotives
35465	x	**CC**	SH	BR 70000
35468	x	**M**	YK	National Railway Museum locomotives
35470	v	**CH**	TM	Tyseley Locomotive Works-based locos
35476	x	**M**	SK	LMS 46233
35479	v	**M**	SH	LNER 61306
35486	x	**M**	TN	LNER 60009/61994

CORRIDOR BRAKE FIRST

Mark 2C. Pressure ventilated. Renumbered when declassified. B4 bogies. ETS 4.

Lot No. 30796 Derby 1969–70. 32.5 t.

35508 (14128, 17128)	**M**	BQ	LMS 44871/45407

CORRIDOR BRAKE FIRST

Mark 2A. Pressure ventilated. Renumbered when declassified. B4 bogies. ETS 4.

Lot No. 30786 Derby 1968. 32 t.

35517 (14088, 17088)	b	**M**	BQ	LMS 44871/45407
35518 (14097, 17097)	b	**G**	SH	SR 34067

COURIER VEHICLE

Mark 1. Converted 1986–87 from Corridor Brake Standards. ETS 2.

80204/217. Lot No. 30699 Wolverton 1962. Commonwealth bogies. 37 t.
80220. Lot No. 30573 Gloucester 1960. B4 bogies. 33 t.

80204 (35297)	**M**	CS	WCRC Carnforth-based locomotives
80217 (35299)	**M**	CS	WCRC Carnforth-based locomotives
80220 (35276)	**M**	NY	LNER 62005

6. 95xxx & 99xxx RANGE NUMBER CONVERSION TABLE

The following table is presented to help readers identify carriages which may still carry numbers in the 95xxx and 99xxx number ranges of the former private owner number series, which is no longer in general use.

9xxxx	BR No.	9xxxx	BR No.	9xxxx	BR No.
95402	Pullman 326	99350	Pullman 350	99673	550
99035	35322	99351	Pullman 351	99674	551
99040	21232	99352	Pullman 352	99675	552
99041	35476	99353	Pullman 353	99676	553
99052	Saloon 41	99354	Pullman 354	99677	586
99121	3105	99361	Pullman 335	99678	504
99122	3106	99371	3128	99679	506
99125	3113	99405	35486	99680	17102
99127	3117	99530	Pullman 301	99710	18767
99128	3130	99531	Pullman 302	99716 *	18808
99131	Saloon 1999	99532	Pullman 308	99718	18862
99241	35494	99534	Pullman 245	99721	18756
99302	13323	99535	Pullman 213	99723	35459
99304	21256	99536	Pullman 254	99880	Saloon 159
99311	1882	99537	Pullman 280	99881	Saloon 807
99312	35463	99539	Pullman 255	99883	2108
99319	17168	99541	Pullman 243	99885	2110
99326	4954	99543	Pullman 284	99887	2127
99327	5044	99546	Pullman 281	99953	35468
99328	5033	99547	Pullman 292	99966	34525
99329	4931	99548	Pullman 293	99970	Pullman 232
99347	Pullman 347	99670	546	99972	Pullman 318
99348	Pullman 348	99671	548	99973	324
99349	Pullman 349	99672	549	99974	Pullman 328

* The number 99716 has also been applied to 3416 for filming purposes.

▲ Chiltern Mainline-liveried Mark 3A Open Standard 12614 (rebuilt with sliding plug doors) is seen near Hatton on 08/04/15. **Robert Pritchard**

▼ Pullman Car Company-liveried Corridor Brake First 17080 at Ravenglass on 25/07/15. **Andrew Mason**

▲ Revised Direct Rail Services-liveried Mark 2D Corridor Brake First 17159 near St Bees on 01/08/15. Note the window bars on this, and adjacent Open Standard 6173, for operating on the Cumbrian Coast. **Robert Pritchard**

▼ BR Carmine & Cream-liveried Mark 1 Corridor Brake Composite 21249 (support coach for steam locomotive 60163) at Goodrington on 02/08/15. **Robin Ralston**

▲ Network Rail yellow-liveried Driving Trailer Coach 82111 leads a test train through Rugby on 02/10/14 powered by 67003 (behind 82111). **Robert Pritchard**

▼ Virgin Trains East Coast-liveried Mark 4 Driving Brake Van 82219 leads the 11.00 Edinburgh–King's Cross through Potters Bar on 05/07/15. **Robert Pritchard**

▲ Stagecoach East Midlands Trains-liveried HST Trailer Kitchen Buffet First 40741 at Nottingham on 03/08/14. **Robert Pritchard**

▼ Virgin Trains East Coast-liveried HST Trailer First 41159 near Hadley Wood on 05/07/15. **Robert Pritchard**

▲ CrossCountry-liveried HST Trailer Standard 42379 at Sheffield on 10/06/15.
Robert Pritchard

▼ First Great Western-liveried HST Trailer Guard's Standard 44076 at Torquay on 08/08/15.
Robin Ralston

▲ Royal Scotsman-liveried Mark 3A Service Car 99969 at Westbury on 15/07/14. **Stewart Armstrong**

▼ Revised DRS-liveried Driving Open Brake Standard 9707 arrives at St Bees with the 14.33 Carlisle–Barrow-in-Furness (powered by 37423) on 08/08/15.
Robert Pritchard

▲ Arlington Fleet Services-liveried EMU Translator Vehicle 975978 at Milton Keynes Central on 18/07/11. **Mark Beal**

▼ Network Rail Mark 2 Structure Gauging Train Coach 99666 at Rugby on 02/10/14. **Robert Pritchard**

▲ Network Rail Mark 2 Structure Gauging Train Coach 977985 near Sheffield on 12/08/15. **Robert Pritchard**

▼ Network Rail Ultrasonic Test Coach 999602 (converted from a Class 421 EMU MBSO) near Market Rasen on 29/12/14. **Robert Pritchard**

7. SET FORMATIONS

CHILTERN RAILWAYS MARK 3 SET FORMATIONS

Chiltern Railways uses a number of loco-hauled rakes on services principally between London Marylebone and Birmingham Moor Street. Rake AL06 is used on a commuter train between Marylebone and Banbury. All coaches apart from rake AL06 and spare 12094 have been rebuilt at Wabtec, Doncaster and fitted with sliding plug doors.

Set							*DVT*
AL01	12610	12613	12614	12615	12602	10273	82301
AL02	12603	12606	12607	12608	12609	10272	82302
AL03	12616	12617	12618	12619	12604	10274	82303
AL04	12623	12605	12621	12627	12625	10271	82304
AL06	12043	12119	12017	11029	11031	12054	82305

Spare 12094 12620 82309

VIRGIN TRAINS EAST COAST MARK 4 SET FORMATIONS

The Virgin Trains East Coast Mark 4 sets generally run in fixed formations since their refurbishment at Bombardier, Wakefield in 2003–05. These rakes are listed below. Class 91 locomotives are positioned next to Coach B.

Set	B	C	D	E	F	H	K	L	M	DVT
BN01	12207	12417	12415	12414	12307	10307	11298	11301	11401	82207
BN02	12232	12402	12450	12448	12302	10302	11299	11302	11402	82202
BN03	12201	12401	12459	12478	12301	10320	11277	11303	11403	82219
BN04	12202	12480	12421	12518	12327	10303	11278	11304	11404	82209
BN05	12209	12486	12520	12522	12300	10326	11219	11305	11405	82210
BN06	12208	12406	12420	12422	12313	10309	11279	11306	11406	82208
BN07	12231	12411	12405	12489	12329	10323	11280	11307	11407	82224
BN08	12205	12481	12485	12407	12328	10300	11229	11308	11408	82211
BN09	12230	12513	12483	12514	12308	10304	11281	11309	11409	82215
BN10	12214	12419	12488	12443	12305	10331	11282	11310	11410	82205
BN11	12203	12437	12436	12484	12315	10308	11283	11311	11411	82218
BN12	12212	12431	12404	12426	12330	10333	11284	11312	11412	82212
BN13	12228	12469	12430	12424	12311	10313	11285	11313	11413	82213
BN14	12229	12410	12526	12423	12312	10332	11201	11314	11414	82206
BN15	12226	12442	12409	12515	12309	10306	11286	11315	11415	82214
BN16	12213	12428	12445	12433	12304	10315	11287	11316	11416	82222
BN17	12223	12444	12427	12432	12303	10324	11288	11317	11417	82225
BN18	12215	12453	12468	12467	12324	10305	11289	11318	11418	82220
BN19	12211	12434	12400	12470	12310	10318	11290	11319	11419	82201
BN20	12224	12477	12439	12440	12326	10321	11241	11320	11420	82200
BN21	12222	12441	12461	12476	12323	10330	11244	11321	11421	82227
BN22	12210	12452	12460	12473	12316	10301	11291	11322	11422	82230
BN23	12225	12454	12456	12455	12318	10325	11292	11323	11423	82226
BN24	12219	12447	12425	12403	12319	10328	11293	11324	11424	82229
BN25	12217	12446	12519	12464	12322	10312	11294	11325	11425	82216
BN26	12220	12474	12465	12429	12325	10311	11295	11326	11426	82223
BN27	12216	12449	12466	12538	12317	10319	11237	11327	11427	82228
BN28	12218	12458	12463	12533	12320	10310	11273	11328	11428	82217
BN29	12204	12462	12457	12438	12321	10317	11998	11329	11429	82231
BN30	12227	12471	12534	12472	12331	10329	11999	11330	11430	82203
Spare	12200									82224

PLATFORM 5 MAIL ORDER

BRITISH RAIL COACHING STOCK VOLUME 3: MK1 NON-PASSENGER CARRYING COACHING STOCK
Coorlea Publishing

The third volume of this reference work giving details of all hauled coaching stock to have operated on the national railway network. This volume covers all non-passenger carrying Mark 1 coaches. Includes details of numbering, conversions, refurbishments and exports. Vehicles are listed in numeric order with previous numbers shown in brackets. Other details provided include current owner where applicable and details of vehicle renumbering, disposal or preservation. 52 pages. **£12.95**

Also available in the same series:
British Rail Coaching Stock Volume 1: Mk 2, 3 & 4 Coaches £11.95
British Rail Coaching Stock Volume 2: Mk1 Passenger Coaches... £16.95

Please add postage: 10% UK, 20% Europe, 30% Rest of World.

Telephone, fax or send your order to the Platform 5 Mail Order Department. See inside back cover of this book for details.

8. SERVICE STOCK

Carriages in this section are used for internal purposes within the railway industry, ie they do not generate revenue from outside the industry. Most are numbered in the former BR departmental number series.

BARRIER, ESCORT & TRANSLATOR VEHICLES

These vehicles are used to move multiple unit, HST and other vehicles around the national railway system.

HST Barrier Vehicles. Mark 1/2A. Renumbered from BR departmental series, or converted from various types. B4 bogies (* Commonwealth bogies).

Non-standard livery: 6340, 6344, 6346 All over dark blue.

6330. Mark 2A. Lot No. 30786 Derby 1968.
6336/38/44. Mark 1. Lot No. 30715 Gloucester 1962.
6340. Mark 1. Lot No. 30669 Swindon 1962.
6346. Mark 2A. Lot No. 30777 Derby 1967.
6348. Mark 1. Lot No. 30163 Pressed Steel 1957.

6330	(14084, 975629)		**FB**	A	*GW*	LA
6336	(81591, 92185)		**FB**	A	*GW*	LA
6338	(81581, 92180)		**FB**	A	*GW*	LA
6340	(21251, 975678)	*	**0**	A	*VE*	EC
6344	(81263, 92080)		**0**	A	*VE*	EC
6346	(9422)		**0**	A	*VE*	EC
6348	(81233, 92963)		**FB**	A	*GW*	LA

Mark 4 Barrier Vehicles. Mark 2A/2C. Converted from Corridor First (*) or Open Brake Standard. B4 bogies.

Non-standard livery: 6358 All over dark blue.

6352/53. Mark 2A. Lot No. 30774 Derby 1968.
6354/55. Mark 2C. Lot No. 30820 Derby 1970.
6358/59. Mark 2A. Lot No. 30788 Derby 1968.

6352	(13465, 19465)	*	**GN**	E	*VE*	BN
6353	(13478, 19478)	*	**GN**	E	*VE*	BN
6354	(9459)		**GN**	E	*VE*	BN
6355	(9477)		**GN**	E	*VE*	BN
6358	(9432)		**0**	E	*VE*	BN
6359	(9429)		**GN**	E	*VE*	BN

EMU Translator Vehicles. Mark 1. Converted 1980 from Restaurant Unclassified Opens. 6376/77 have Tighlock couplers and 6378/79 Dellner couplers. Commonwealth bogies.

Lot No. 30647 Wolverton 1959–61.

6376	(1021, 975973)	**PB**	P	*CS*	ZG	*(works with 6377)*
6377	(1042, 975975)	**PB**	P	*CS*	ZG	*(works with 6376)*
6378	(1054, 975971)	**PB**	P	*CS*	ZG	*(works with 6379)*
6379	(1059, 975972)	**PB**	P	*CS*	ZG	*(works with 6378)*

SERVICE STOCK 101

HST Barrier Vehicles. Mark 1. Converted from Gangwayed Brake Vans in 1994–95. B4 bogies.

6392. Lot No. 30715 Gloucester 1962.
6393/97. Lot No. 30716 Gloucester 1962.
6394. Lot No. 30162 Pressed Steel 1956–57.
6398/99. Lot No. 30400 Pressed Steel 1957–58.

6392	(81588, 92183)	**PB**	P		LM
6393	(81609, 92196)	**PB**	P	*VE*	EC
6394	(80878, 92906)	**P**	P	*VE*	EC
6397	(81600, 92190)	**PB**	P		LM
6398	(81471, 92126)	**PB**	EM	*EM*	NL
6399	(81367, 92994)	**PB**	EM	*EM*	NL

Escort Coaches. Converted from Mark 2A Open Brake Standards. These vehicles use the same bodyshell as the Mark 2A Corridor Brake First. B4 bogies.

9419. Lot No.30777 Derby 1970.
9428. Lot No.30820 Derby 1970.

9419	**DS**	DR	*DR*	KM
9428	**DS**	DR	*DR*	KM

EMU Translator Vehicles. Converted from Class 508 driving cars.

64664. Lot No. 30979 York 1979–80.
64707. Lot No. 30981 York 1979–80.

64664	**AG**	A	*GB*	ZG	James D. Rowlands	*(works with 64707)*
64707	**AG**	A	*GB*	ZG	Sir David Rowlands	*(works with 64664)*

Eurostar Barrier Vehicles. Mark 1. Converted from General Utility Vans with bodies removed. Fitted with B4 bogies for use as Eurostar barrier vehicles.

96380/381. Lot No. 30417 Pressed Steel 1958–59.
96383. Lot No. 30565 Pressed Steel 1959.
96384. Lot No. 30616 Pressed Steel 1959–60.

96380	(86386, 6380)	**B**	EU	*EU*	TI
96381	(86187, 6381)	**B**	EU	*EU*	TI
96383	(86664, 6383)	**B**	EU	*EU*	TI
96384	(86955, 6384)	**B**	EU	*EU*	TI

EMU Translator Vehicles. Converted from various Mark 1s.

Non-standard livery: All over blue.

975864. Lot No. 30054 Eastleigh 1951–54. Commonwealth bogies.
975867. Lot No. 30014 York 1950–51. Commonwealth bogies.
975875. Lot No. 30143 Charles Roberts 1954–55. Commonwealth bogies.
975974/978. Lot No. 30647 Wolverton 1959–61. B4 bogies.
977087. Lot No. 30229 Metro–Cammell 1955–57. Commonwealth bogies.

975864	(3849)	**HB**	E	*GB*	PG	*(works with 975867)*
975867	(1006)	**HB**	E	*GB*	PG	*(works with 975864)*
975875	(34643)	**0**	E	*GB*	PG	*(works with 977087)*

SERVICE STOCK

975974	(1030)	**AG**	A	*GB*	ZG	Paschar	*(works with 975978)*
975978	(1025)	**AG**	A	*GB*	ZG	Perpetiel	*(works with 975974)*
977087	(34971)	**O**	E	*GB*	PG		*(works with 975875)*

LABORATORY, TESTING & INSPECTION COACHES

These coaches are used for research, development, testing and inspection on the national railway system. Many are fitted with sophisticated technical equipment.

Plain Line Pattern Recognition Coaches. Converted from BR Mark 2F Buffet First (*) or Open Standard. B4 bogies.

1256. Lot No. 30845 Derby 1973.
5981. Lot No. 30860 Derby 1973–74.

1256	(3296)	*	**Y**	NR	*CS*	ZA
5981			**Y**	NR	*CS*	ZA

Generator Vans. Mark 1. Converted from BR Mark 1 Gangwayed Brake Vans. B5 bogies.

6260. Lot No. 30400 Pressed Steel 1957–58.
6261. Lot No. 30323 Pressed Steel 1957.
6262. Lot No. 30228 Metro-Cammell 1957–58.
6263. Lot No. 30163 Pressed Steel 1957.
6264. Lot No. 30173 York 1956.

6260	(81450, 92116)	**Y**	NR	*CS*	ZA
6261	(81284, 92988)	**Y**	NR	*CS*	ZA
6262	(81064, 92928)	**Y**	NR	*CS*	ZA
6263	(81231, 92961)	**Y**	NR	*CS*	ZA
6264	(80971, 92923)	**Y**	NR	*CS*	ZA

Staff Coach. Mark 2D. Converted from BR Mark 2D Open Brake Standard. Lot No. 30824 Derby 1971. B4 bogies.

9481	**Y**	NR	*CS*	ZA

Test Train Brake Force Runners. Mark 2F. Converted from BR Mark 2F Open Brake Standard. Lot No. 30861 Derby 1974. B4 bogies.

9516	**Y**	NR	*CS*	ZA
9523	**Y**	NR	*CS*	ZA

Driving Trailer Coaches. Converted 2008 at Serco, Derby from Mark 2F Driving Open Brake Standards. Fitted with generator. Disc brakes. B4 bogies.

9701–08. Lot No. 30861 Derby 1974. Converted to Driving Open Brake Standard Glasgow 1974.
9714. Lot No. 30861 Derby 1974. Converted to Driving Open Brake Standard Glasgow 1986.

9701	(9528)	**Y**	NR	*CS*	ZA
9702	(9510)	**Y**	NR	*CS*	ZA
9703	(9517)	**Y**	NR	*CS*	ZA
9708	(9530)	**Y**	NR	*CS*	ZA
9714	(9536)	**Y**	NR	*CS*	ZA

SERVICE STOCK

Ultrasonic Test Coach. Converted from Class 421 EMU MBSO.

62287. Lot No. 30808. York 1970. SR Mark 6 bogies.
62384. Lot No. 30816. York 1970. SR Mark 6 bogies.

62287	**Y**	NR	*CS*	ZA
62384	**Y**	NR	*CS*	ZA

Test Train Brake Force Runners. Converted from Mark 2F Open Standard converted to Class 488/3 EMU TSOLH. These vehicles are included in test trains to provide brake force and are not used for any other purposes. Lot No. 30860 Derby 1973–74. B4 bogies.

72612	(6156)	**Y**	NR	*CS*	ZA
72616	(6007)	**Y**	NR	*CS*	ZA

Structure Gauging Train Coach. Converted from Mark 2F Open Standard converted to Class 488/3 EMU TSOLH. Lot No. 30860 Derby 1973–74. B4 bogies.

72630	(6094)	**Y**	NR	*CS*	ZA	*(works with 99666)*

Plain Line Pattern Recognition Coaches. Converted from BR Mark 2F Open Standard converted to Class 488/3 EMU TSOLH. Lot No. 30860 Derby 1973–74. B4 bogies.

72631	(6096)	**Y**	NR	*CS*	ZA
72639	(6070)	**Y**	NR	*CS*	ZA

Driving Trailer Coaches. Converted from Mark 3B 110 mph Driving Brake Vans. Fitted with diesel generator. Lot No. 31042 Derby 1988.

82111	**Y**	NR		ZA
82124	**Y**	NR		ZA
82129	**Y**	NR	*CS*	ZA
82145	**Y**	NR	*CS*	ZA

Structure Gauging Train Coach. Converted from BR Mark 2E Open First then converted to exhibition van. Lot No. 30843 Derby 1972–73. B4 bogies.

99666	(3250)	**Y**	NR	*CS*	ZA	*(works with 72630)*

Inspection Saloon. Converted from Class 202 DEMU TRB at Stewarts Lane for use as a BR Southern Region General Manager's Saloon. Overhauled at FM Rail, Derby in 2004–05 for use as a New Trains Project Saloon. Can be used in push-pull mode with suitably equipped locomotives. Eastleigh 1958. SR Mark 4 bogies.

975025	(60755)	**G**	NR	*CS*	ZA	CAROLINE

Overhead Line Equipment Test Coach ("MENTOR"). Converted from BR Mark 1 Corridor Brake Standard. Lot No. 30142 Gloucester 1954–55. Fitted with pantograph. B4 bogies.

975091	(34615)	**Y**	NR	*CS*	ZA

New Measurement Train Conference Coach. Converted from prototype HST TF Lot No. 30848 Derby 1972. BT10 bogies.

975814	(11000, 41000)	**Y**	NR	*CS*	EC

SERVICE STOCK

New Measurement Train Lecture Coach. Converted from prototype HST catering vehicle. Lot No. 30849 Derby 1972–73. BT10 bogies.

975984 (10000, 40000)   **Y**   NR   *CS*   EC

Radio Survey Coach. Converted from BR Mark 2E Open Standard. Lot No. 30844 Derby 1972–73. B4 bogies.

977868 (5846)   **Y**   NR   *CS*   ZA

Staff Coach. Converted from Royal Household couchette Lot No. 30889, which in turn had been converted from BR Mark 2B Corridor Brake First. Lot No. 30790 Derby 1969. B5 bogies.

977969 (14112, 2906)   **Y**   NR   *CS*   ZA

Track Inspection Train Coach. Converted from BR Mark 2E Open Standard. Lot No. 30844 Derby 1972–73. B4 bogies.

977974 (5854)   **Y**   NR   *CS*   ZA

Electrification Measurement Coach. Converted from BR Mark 2F Open First converted to Class 488/2 EMU TFOH. Lot No. 30859 Derby 1973–74. B4 bogies.

977983 (3407, 72503)   **Y**   NR   *CS*   ZA

New Measurement Train Staff Coach. Converted from HST catering vehicle. Lot No. 30884 Derby 1976–77. BT10 bogies.

977984 (40501)   **Y**   P   *CS*   EC

Structure Gauging Train Coaches. Converted from Mark 2F Open Standard converted to Class 488/3 EMU TSOLH or from BR Mark 2D Open First subsequently declassified to Open Standard and then converted to exhibition van. B4 bogies.
977985. Lot No. 30860 Derby 1973–74.
977986. Lot No. 30821 Derby 1971.

977985 (6019, 72715)   **Y**   NR   *CS*   ZA *(works with 977986)*
977986 (3189, 99664)   **Y**   NR   *CS*   ZA *(works with 977985)*

New Measurement Train Overhead Line Equipment Test Coach. Converted from HST TGS. Lot No. 30949 Derby 1982. Fitted with pantograph. BT10 bogies.

977993 (44053)   **Y**   P   *CS*   EC

New Measurement Train Track Recording Coach. Converted from HST TGS. Lot No. 30949 Derby 1982. BT10 bogies.

977994 (44087)   **Y**   P   *CS*   EC

New Measurement Train Coach. Converted from HST catering vehicle. Lot No. 30921 Derby 1978–79. BT10 bogies. Fitted with generator.

977995 (40719, 40619)   **Y**   P   *CS*   EC

Radio Survey Coach. Converted from Mark 2F Open Standard converted to Class 488/3 EMU TSOLH. Lot No. 30860 Derby 1973–74.

977997 (72613, 6126)   **Y**   NR   *CS*   ZA

SERVICE STOCK

Track Recording Coach. Purpose built Mark 2. B4 bogies.

| 999550 | | Y | NR | *CS* | ZA |

Ultrasonic Test Coaches. Converted from Class 421 EMU MBSO and Class 432 EMU MSO.

999602/605. Lot No. 30862 York 1974. SR Mk 6 bogies.
999606. Lot No. 30816. York 1970. SR Mk 6 bogies.

999602	(62483)	Y	NR	*CS*	ZA
999605	(62482)	Y	NR	*CS*	ZA
999606	(62356)	Y	NR	*CS*	ZA

BREAKDOWN TRAIN COACHES

These coaches are formed in trains used for the recovery of derailed railway vehicles and were converted from BR Mark 1 Corridor Brake Standard and General Utility Van. The current use of each vehicle is given.

971001/003/004. Lot No. 30403 York/Glasgow 1958–60. Commonwealth bogies.
971002. Lot No. 30417 Pressed Steel 1958–59. Commonwealth bogies.
975087. Lot No. 30032 Wolverton 1951–52. BR Mark 1 bogies.
975464. Lot No. 30386 Charles Roberts 1956–58. Commonwealth bogies.
975471. Lot No. 30095 Wolverton 1953–55. Commonwealth bogies.
975477. Lot No. 30233 GRCW 1955–57. BR Mark 1 bogies.
975486. Lot No. 30025 Wolverton 1950–52. Commonwealth bogies.

971001	(86560, 94150)	Y	NR	*DB*	BS	Tool & Generator Van
971002	(86624, 94190)	Y	NR	*DB*	SP	Tool Van
971003	(86596, 94191)	Y	NR	*DB*	BS	Tool Van
971004	(86194, 94168)	Y	NR	*DB*	KY	Tool Van
975087	(34289)	Y	NR	*DB*	KY	Tool & Generator Van
975464	(35171)	Y	NR	*DB*	SP	Staff Coach
975471	(34543)	Y	NR	*DB*	BS	Staff Coach
975477	(35108)	Y	NR	*DB*	KY	Staff Coach
975486	(34100)	Y	NR	*DB*	SP	Tool & Generator Van

INFRASTRUCTURE MAINTENANCE COACHES

De-Icing Coaches

These coaches are used for removing ice from the conductor rail of DC lines. They were converted from Class 489 DMLVs that had originally been Class 414/3 DMBSOs.

Lot No. 30452 Ashford/Eastleigh 1959. Mk 4 bogies.

68501	(61281)	Y	NR	*GB*	ZG
68504	(61286)	Y	NR	*GB*	ZG
68505	(61299)	Y	NR	*GB*	ZG

Winterisation Train Coach. Converted from BR Mark 2E Open Standard. Lot No. 30844 Derby 1972–73. B4 bogies.

| 977869 (5858) | | Y | NR | *DR* | Edinburgh Slateford |

INTERNAL USER VEHICLES

These vehicles are confined to yards and depots or do not normally move at all. Details are given of the internal user number (if allocated), type, former identity, current use and location. Many no longer see regular use.

975403 carries its original number 4598.

024787*	BR GUV 93219	Stores Van	Nottingham Eastcroft Depot
041379	LMS CCT DM 395951	Stores van	Leeman Road EY, York
041474*	BR SPV 975418	Stores Van	Worksop Up Yard
041947	BR GUV 93425	Stores van	Ilford Depot (London)
041989*	BR SPV 975423	Stores Van	Toton Depot
042154	BR GUV 93975	Stores van	Ipswich Upper Yard
061061	BR CCT 94135	Stores van	Oxford Station
061202*	BR GUV 93498	Stores van	Laira Depot (Plymouth)
061223	BR GUV 93714	Stores van	Oxford Station
083602	BR CCT 94494	Stores van	Three Bridges Station
083637	BR NW 99203	Stores van	Stewarts Lane Depot
083644	BR Ferry Van 889201	Stores van	Eastleigh Depot
083664	BR Ferry Van 889203	Stores van	Eastleigh Depot
–	BR Open First 3186	Instruction Coach	Derby Etches Park Depot
–	BR Open First 3381	Instruction Coach	Hornsey Depot (London)
–	BR Open Standard 5636	Instruction Coach	St Philip's Marsh Depot
–	BR BV 6360	Barrier vehicle	Neville Hill Depot, Leeds
–	BR BV 6396	Stores van	Longsight Depot (Manchester)
–	BR RFKB 10256	Instruction Coach	Yoker Depot
–	BR RFKB 10260	Instruction Coach	Yoker Depot
–	BR BFK 17156	Instruction Coach	Derby Etches Park Depot
–	BR SPV 88045*	Stores Van	Thameshaven Yard
–	BR BG 92901	Stores van	Wembley Depot (London)
–	BR NL 94003	Stores van	Old Oak Common HST Depot
–	BR NL 94006	Stores van	Old Oak Common HST Depot
–	BR NK 94121	Stores van	Toton Depot
–	BR NB 94438	Stores van	Toton Depot
–	BR CCT 94663*	Stores Van	Mossend Up Yard
–	BR GUV 96139	Stores van	Longsight Depot (Manchester)
–	BR Ferry Van 889200	Stores van	Stewarts Lane Depot
–	BR Ferry Van 889202	Stores van	Stewarts Lane Depot
–	BR Open Standard 975403	Cinema Coach	St Philip's Marsh Depot
–	SR PMV 977045*	Stores Van	EMD, Longport Works
–	SR CCT 2516*	Stores Van	Eastleigh Depot

* Grounded body

Abbreviations:
BFK = Corridor Brake First
BG = Gangwayed Brake Van
BV = Barrier Vehicle
CCT = Covered Carriage Truck (a 4-wheeled van similar to a GUV)
GUV = General Utility Van (bogied van with side and end doors)
NB = High Security Brake Van (converted from BG, gangways removed)
NK = High Security General Utility Van (end doors removed)

COACHING STOCK AWAITING DISPOSAL 107

NL = Newspaper Van (converted from a GUV)
NW = Bullion Van (converted from a Corridor Brake Standard)
PMV = Parcels & Miscellaneous Van (a 4-wheeled van similar to a CCT but without end doors)
RFKB = Kitchen Buffet First
SPV = Special Parcels Van (a 4-wheeled van converted from a Fish Van)

9. COACHING STOCK AWAITING DISPOSAL

This list contains the last known locations of carriages awaiting disposal. The definition of which vehicles are "awaiting disposal" is somewhat vague, but generally speaking these are vehicles of types not now in normal service, those not expected to see further use or carriages which have been damaged by fire, vandalism or collision.

1252	SH	4849	CS	6029	SH	12144	LM
1253	SH	4854	CS	6041	CS	12156	LM
1258	CS	4860	CS	6045	SH	12160	LM
1644	CS	4932	CS	6050	CS	12163	LM
1650	CS	4997	CS	6073	SH	13306	CS
1652	CS	5148	TM	6134	SH	13323	CS
1655	CS	5179	TM	6151	SH	13508	BU
1658	CL	5183	TM	6154	SH	17168	CS
1663	CS	5186	TM	6175	CS	18767	SH
1670	CS	5193	TM	6179	CS	18808	SH
1679	CL	5194	TM	6324	CP	18862	SH
1680	CL	5221	TM	6361	NL	34525	CS
1696	CL	5331	FA	6364	WH	40729	NL
1800	SH	5386	FA	6365	WH	41043	LB
2108	CS	5420	TM	6412	BU	80212	CS
2110	CS	5453	CS	6720	FA	80403	CS
2127	CS	5463	CS	9440	SH	80404	CS
2131	CS	5478	CS	9489	CS	80414	SL
2833	CS	5491	CS	10201	LM	82125	LM
2834	EH	5569	CS	10245	CS	84519	BU
2909	CS	5737	CS	10530	ZH	92114	ZA
3241	CS	5740	CS	10588	ZH	92159	CS
3255	FA	5756	CS	10682	LM	92303	CL
3303	TO	5815	SH	11005	LM	92400	BU
3309	CS	5876	SH	11021	LM	92908	CS
3368	FA	5888	SH	12008	LM	92936	CL
3388	FA	5925	SH	12029	LM	93723	BY
3399	FA	5943	CS	12036	LM	94101	CS
3408	CS	5958	SH	12095	LM	94104	TO
3416	CS	5978	SH	12096	ZN	94106	BU
4362	BU	6009	SH	12101	LM	94116	BU

COACHING STOCK AWAITING DISPOSAL

94153	WE	94344	TO	94527	Hellifield	96604	CF
94160	BO	94401	CS	94530	CS	96605	RU
94166	BS	94406	CS	94531	BU	96606	RU
94170	CL	94408	CS	94538	CL	96607	RU
94176	BS	94410	WE	94539	CS	96608	RU
94177	TO	94420	CS	94540	TJ	96609	RU
94195	BS	94422	TO	94542	Hellifield	99019	CS
94196	CS	94423	BS	94544	BO	99884	CS
94197	BS	94427	WE	94545	Tees Yard	889400	ZA
94199	BO	94428	CS	94546	Hellifield	975081	ZA
94207	TO	94429	Tees Yard	94547	CS	975280	ZA
94208	TO	94431	CS	94548	CS	975454	TO
94214	CS	94434	BU	95300	CS	975484	CS
94217	BO	94435	TO	95301	CS	975639	CS
94222	CS	94445	WE	95410	CS	975681	Portobello
94227	Tees Yard	94450	WE	95727	WE	975682	Portobello
94229	CL	94451	WE	95754	CS	975685	Portobello
94302	Hellifield	94470	TO	95761	WE	975686	Portobello
94303	Hellifield	94479	TO	95763	BS	975687	Portobello
94304	MH	94482	CS	96110	CS	975688	Portobello
94306	Hellifield	94488	CL	96132	CS	975920	Portobello
94308	CS	94490	BU	96135	CS	977077	**
94310	WE	94492	WE	96164	CS	977085	BU
94311	WE	94495	Hellifield	96165	CS	977095	CS
94313	WE	94497	BO	96170	CS	977111	**
94316	TO	94498	CS	96178	CS	977112	**
94317	TO	94501	TO	96182	CS	977169	BU
94322	CS	94504	Hellifield	96191	CS	977618	BY
94323	Hellifield	94512	CS	96192	CS	977989	ZA
94326	Hellifield	94514	BO	96371	WB		
94332	CS	94515	**	96372	LM	DS 70220	**
94333	Hellifield	94517	CL	96373	LM		
94335	BU	94520	BU	96374	ZB	Pullman 315	CS
94336	CL	94522	BU	96375	LM	Pullman 316	CS
94337	WE	94525	Hellifield	96602	RU	Pullman 325	CS
94338	WE	94526	Hellifield	96603	CF	Pullman 337	CS

** Other locations:

94515	Eastleigh East Yard
977077	Ripple Lane Yard
977111	Ripple Lane Yard
977112	Ripple Lane Yard
DS 70220	Western Trading Estate Siding, North Acton

10. CODES

10.1. LIVERY CODES

The colour of the lower half of the bodyside is stated first. Minor variations to these liveries are ignored.

1	"One" (metallic grey with a broad black bodyside stripe. White National Express/Abellio Greater Anglia "interim" stripe as branding).
AG	Arlington Fleet Services (green).
AL	Advertising/promotional livery (see class heading for details).
AR	Anglia Railways (turquoise blue with a white stripe).
AV	Arriva Trains (turquoise blue with white doors).
AW	Arriva Trains Wales/Welsh Government sponsored dark & light blue.
B	BR blue.
BG	BR blue & grey lined out in white.
BP	Blue Pullman ("Nanking" blue & white).
CA	Caledonian Sleeper (dark blue).
CC	BR Carmine & Cream.
CH	BR Western Region/GWR chocolate & cream lined out in gold.
CM	Chiltern Mainline (two-tone grey & silver with blue stripes).
DR	Direct Rail Services (dark blue with light blue or dark grey roof).
DS	Revised Direct Rail Services (dark blue, light blue & green. "Compass" logo).
EC	East Coast (silver or grey with a purple stripe).
FB	First Group dark blue.
FD	First Great Western "Dynamic Lines" (dark blue with thin multi-coloured lines on lower bodyside).
FP	Old First Great Western (green & ivory with thin green & broad gold stripes).
FS	First Group (indigo blue with pink & white stripes).
G	BR Southern Region/SR green.
GA	Abellio Greater Anglia (white with red doors & black window surrounds).
GC	Grand Central (black with an orange stripe).
GN	Great North Eastern Railway {modified} (dark blue with a white stripe).
GW	Great Western Railway (dark green).
HB	HSBC Rail (Oxford blue & white)
IC	BR InterCity (light grey/red stripe/white stripe/dark grey).
M	BR maroon (maroon lined out in straw & black).
NX	National Express (white with grey ends).
O	Non-standard (see class heading for details).
P	Porterbrook Leasing Company (purple & grey or white).
PB	Porterbrook Leasing Company blue.
PC	Pullman Car Company (umber & cream with gold lettering lined out in gold).
RB	Riviera Trains Oxford blue.
RP	Royal Train (claret, lined out in red & black).
RV	Riviera Trains Great Briton (Oxford blue & cream, lined out in gold).

SR	ScotRail – Scotland's Railways (dark blue with Scottish Saltire flag & white/blue flashes).
ST	Stagecoach {long-distance stock} (white & dark blue with dark blue window surrounds and red & orange swishes at unit ends).
U	White or grey undercoat.
V	Virgin Trains (red with black doors extending into bodysides, three white lower bodyside stripes).
VE	Virgin Trains East Coast (red & white with black window surrounds).
VN	Belmond Northern Belle (crimson lake & cream lined out in gold).
VT	Virgin Trains silver (silver, with black window surrounds, white cantrail stripe and red roof. Red swept down at unit ends).
XC	CrossCountry (two tone silver with deep crimson ends & pink doors).
Y	Network Rail yellow.

10.2. OWNER CODES

The following codes are used to define the ownership details of the rolling stock listed in this book. Codes shown indicate either the legal owner or "responsible custodian" of each vehicle.

A	Angel Trains
AV	Arriva UK Trains
BE	Belmond (UK)
DB	DB Schenker Rail (UK)
DR	Direct Rail Services
E	Eversholt Rail (UK)
EM	East Midlands Trains
EU	Eurostar (UK)
FG	First Group
LS	Locomotive Services
NR	Network Rail
NY	North Yorkshire Moors Railway Enterprises
P	Porterbrook Leasing Company
PR	The Princess Royal Class Locomotive Trust
RF	Railfilms
RP	Rampart Engineering
RV	Riviera Trains
SP	The Scottish Railway Preservation Society
VT	Vintage Trains
WC	West Coast Railway Company

10.3. OPERATOR CODES

The two letter operator codes give the current operator (by trading name). This is the organisation which facilitates the use of the coach and may not be the actual Train Operating Company which runs the train on which the particular coach is used. If no operator code is shown then the vehicle is not in use at present.

AW Arriva Trains Wales
BP Belmond British Pullman
CA Caledonian Sleeper
CR Chiltern Railways
CS Colas Rail
DB DB Schenker
DR Direct Rail Services
EM East Midlands Trains
EU Eurostar (UK)
GA Abellio Greater Anglia
GB GB Railfreight
GC Grand Central
GW Great Western Railway
LS Locomotive Services
NB Belmond Northern Belle
NO Northern
NY North Yorkshire Moors Railway
PR The Princess Royal Class Locomotive Trust
RS The Royal Scotsman (Belmond)
RT Royal Train
RV Riviera Trains
SP The Scottish Railway Preservation Society
SR ScotRail
ST Statesman Rail
VE Virgin Trains East Coast
VT Vintage Trains
WC West Coast Railway Company
XC CrossCountry

10.4. ALLOCATION & LOCATION CODES

Code	Depot	Operator
AL	Aylesbury	Chiltern Railways
BH	Barrow Hill (Chesterfield)	Barrow Hill Engine Shed Society
BN	Bounds Green (London)	Virgin Trains East Coast
BO	Bo'ness (West Lothian)	The Bo'ness & Kinneil Railway
BQ	Bury (Greater Manchester)	East Lancashire Rly Trust/Riley & Son (Railways)
BS	Bescot (Walsall)	DB Schenker Rail (UK)
BU	Burton-upon-Trent	Nemesis Rail
BY	Bletchley	London Midland
CL	Crewe LNWR Heritage	LNWR Heritage Company

Code	Location	Operator
CF	Cardiff Canton	Arriva Trains Wales/Colas Rail
CM	East Cranmore	Cranmore Railway Company
CP	Crewe Arriva Traincare	Arriva Traincare
CS	Carnforth	West Coast Railway Company
CY	Crewe South Yard	*Storage location only*
EC	Edinburgh Craigentinny	Virgin Trains East Coast
EH	Eastleigh Arriva Traincare	Arriva Traincare
FA	Fawley (Hampshire)	*Storage location only*
IL	Ilford (London)	Abellio Greater Anglia
KM	Carlisle Kingmoor	Direct Rail Services
KY	Knottingley	DB Schenker Rail (UK)
LA	Laira (Plymouth)	Great Western Railway
LB	Loughborough Works	Wabtec Rail
LM	Long Marston (Warwickshire)	Quinton Rail Technology Centre
MH	Millerhill Yard (Edinburgh)	DB Schenker Rail (UK)
ML	Motherwell	Direct Rail Services
NC	Norwich Crown Point	Abellio Greater Anglia
NL	Neville Hill (Leeds)	East Midlands Trains/Northern
NY	Grosmont (North Yorkshire)	North Yorkshire Moors Railway Enterprises
OO	Old Oak Common HST	Great Western Railway
PG	Peterborough GBRf	GB Railfreight
PM	St Philip's Marsh (Bristol)	Great Western Railway
PO	Polmadie (Glasgow)	Alstom
PZ	Penzance Long Rock	Great Western Railway
RU	Rugby	Colas Rail
SH	Southall (London)	West Coast Railway Co/Locomotive Services
SK	Swanwick Junction (Derbyshire)	Midland Railway Enterprises
SL	Stewarts Lane (London)	Govia Thameslink Railway/Belmond
SP	Springs Branch (Wigan)	DB Schenker Rail (UK)
TJ	Tavistock Junction Yard (Plymouth)	*Storage location only*
TI	Temple Mills (London)	Eurostar (UK)
TM	Tyseley Locomotive Works	Birmingham Railway Museum
TN	Thornton (Fife)	John Cameron
TO	Toton (Nottinghamshire)	DB Schenker Rail (UK)
WB	Wembley (London)	Alstom
WE	Willesden Brent Sidings	*Storage location only*
WH	Washwood Heath (Birmingham)	Boden Rail Engineering
YK	National Railway Museum (York)	National Museum of Science & Industry
ZA	RTC Business Park (Derby)	Railway Vehicle Engineering
ZB	Doncaster Works	Wabtec Rail
ZC	Crewe Works	Bombardier Transportation UK
ZD	Derby Works	Bombardier Transportation UK
ZG	Eastleigh Works	Arlington Fleet Services
ZH	Springburn Depot (Glasgow)	Knorr-Bremse Rail Systems (UK)
ZI	Ilford Works	Bombardier Transportation UK
ZJ	Stoke-on-Trent Works	Axiom Rail (Stoke)
ZK	Kilmarnock Works	Wabtec Rail Scotland
ZN	Wolverton Works	Knorr-Bremse Rail Systems (UK)
ZR	York (Holgate Works)	Network Rail